Salters-Nuffield Advanced Biology

A2

www.heinemann.co.uk
✓ Free online support
✓ Useful weblinks
✓ 24 hour online ordering

01865 888058

Heinemann

Heinemann Educational Publishers
Halley Court, Jordan Hill, Oxford OX2 8EJ
Part of Harcourt Education

Heinemann is the registered trademark of
Harcourt Education Limited

© University of York Science Education Group 2006

First published 2006
Published as trial edition 2003

10 09 08 07 06
10 9 8 7 6 5 4 3 2

British Library Cataloguing in Publication Data is available
from the British Library on request.

10-digit ISBN: 0 435628 58 5

13-digit ISBN: 978 0 435628 58 1

Project editors: Angela Hall, Michael Reiss, Catherine Rowell and Anne Scott

Heinemann editorial team: Sonia Clark, Andrew Halcro-Johnston

Edited by Ruth Holmes

Index compiled by Laurence Errington

Designed and typeset by Bridge Creative Services Limited, Bicester, Oxon

Original illustrations © Harcourt Education Limited 2006

Illustrated by Roger Farrington and Hardlines Ltd

Printed and bound in Italy by Printer Trento, S.r.l.

For text and photograph acknowledgements, please see page 265.

Contents

Contributors

Many people from schools, colleges, universities, industries and the professions have contributed to the Salters-Nuffield Advanced Biology project. They include the following.

Central team

Angela Hall (Project Officer), Nuffield Curriculum Centre
Michael Reiss (Director), Institute of Education, University of London
Catherine Rowell (Project Officer), University of York Science Education Group
Anne Scott (Project Officer), University of York Science Education Group

Sarah Codrington, Nuffield Curriculum Centre
Nancy Newton (Secretary), University of York Science Education Group

Advisory committee

Professor R McNeill Alexander FRS   University of Leeds
Dr Roger Barker   University of Cambridge
Dr Allan Baxter   GlaxoSmithKline
Professor Sir Tom Blundell FRS (Chair)   University of Cambridge
Professor Kay Davies CBE FRS   University of Oxford
Professor Sir John Krebs FRS   Food Standards Agency
Professor John Lawton FRS   Natural Environment Research Council
Professor Peter Lillford CBE   University of York
Dr Roger Lock   University of Birmingham
Professor Angela McFarlane   University of Bristol
Dr Alan Munro   University of Cambridge
Professor Lord Robert Winston   Imperial College of Science, Technology and Medicine

Authors

Topic 5
Jamie Copsey   Jersey Zoo
Malcolm Ingram
Ron Pickering   Sir John Deane's College, Northwich
David Slingsby   Wakefield Girls' High School
Mark Smith   Leeds Grammar School
Jane Wilson   Eggbuckland Community College, Plymouth

Topic 6
Jon Duveen   City & Islington College, London
Brian Ford   The Sixth Form College, Colchester
Richard Fosbery   The Skinners' School, Tunbridge Wells
Pauline Lowrie   Sir John Deane's College, Northwich
Liz Jackson   King James's School, Knaresborough
Christine Knight

Topic 7
David Greenwood   Greenhead College, Huddersfield
Nick Owens   Oundle School, Peterborough
Jacqui Punter   Brighton, Hove & Sussex Sixth Form College
Catherine Rowell   University of York

Topic 8
Nan Davis
Ginny Hales   Cambridge Regional College
Steve Hall   King Edward VI School, Southampton
Gill Hickman   Ringwood School
Jenny Owens   Rye St Antony School, Oxford
Mark Winterbottom   King Edward VI School, Bury St Edmunds

We would also like to thank the following for their advice and assistance:
Judith Bennett   University of York
Bob Campbell   University of York
Valerie Corrigal   Kings College, London
Rachel Hadi-Talab   Institute of Education, University of London
John Holman   University of York
Andrew Hunt   Nuffield Curriculum Centre
Jenny Lewis   University of Leeds

Sponsors

The Salters' Institute   Pfizer Limited
The Nuffield Foundation   Boots plc
The Wellcome Trust   ICI plc
Zeneca Agrochemicals   The Royal Society of Chemistry

About the SNAB course

Welcome to the second year of the Salters-Nuffield Advanced Biology course (SNAB). SNAB is much more than just another A-level specification. It is a complete course with its own distinctive philosophy and it is supported by exciting teaching, learning and support materials. SNAB combines the key concepts underpinning biology today with the opportunity to gain the wider skills that biologists now need.

SNAB is the result of three years of piloting, funded by a number of organisations including the Salters' Institute and the Nuffield Foundation. Following the pilot, we have been able to incorporate student and teacher feedback into the course to make it even better.

A context-led approach

All eight topics use a context-led approach with a storyline or contemporary issue presented and biological principles introduced when required to aid understanding of the context. In each topic you will study more than one area of biology. In later topics, you will meet many of the ideas again and develop them further. For example, immunity appears in Topic 6 Infection, immunity and forensics, and is revisited in Topic 7 Run for your life, in the context of overtraining.

Building knowledge through the course

In SNAB there is not, for example, a topic labelled 'biochemistry' containing everything you might need to know on carbohydrates, fats, nucleic acids and proteins. In SNAB you study the biochemistry of these large molecules bit by bit throughout the course when you need to know the relevant information for a particular topic. In this way information is presented in manageable chunks and builds on existing knowledge. Ideas introduced in AS are revisited in A2. For example, the control of heart rate in Topic 7 extends the work done on heart and circulation in the first AS topic.

Activities as an integral part of the learning process

You really have to get actively involved in SNAB. Throughout this book you will find references to a wide variety of activities. Through these you will be learning practical and other techniques as well as developing a wide range of skills including, for instance, data analysis, critical evaluation of information, communication and collaborative working.

Within the electronic resources you will find animations on such things as photosynthesis, muscle contraction and transmission of the nerve impulse. These animations are designed to help you understand the more difficult bits of biology. The support sections will be useful if you need help with biochemistry, mathematics and statistics, ICT, study skills, the examination or coursework.

SNAB and ethical debate

With rapid developments in biological science, we are faced with an increasing number of challenging decisions. The use of drugs in sport is one ethical dilemma considered in the A2 course.

In SNAB you develop the ability to discuss and debate these types of biological issues. There is rarely a right or wrong answer; but you learn to justify your own decisions using ethical frameworks.

Exams and coursework

The exams in SNAB are much like any advanced biology exams but particularly reward your ability to reason biologically and to use what you have learned in new contexts. You won't just be repeating information you have learned off by heart. Most of the exam questions are structured ones, though as you go through the course you begin to do short essays, building up to longer ones. We believe that essay writing will be very useful for you if you go on to university or to any sort of job that requires you to be able to write reports. During the second year of the course you will spend two weeks on a coursework investigation of your own choice. You can find out more about the coursework and examinations in the electronic exam and coursework support section and in the specification.

We think that SNAB is the most exciting and up-to-date advanced biology course around. Whatever your interests are – whether you want to study a biological subject at university or use the qualification in other ways – we hope you enjoy the course.

Any questions or comments?

You can write to us at:

The Salters-Nuffield Advanced Biology Project
Science Education Group
University of York
Heslington
York   YO10 5DD
Email: uyseg-snab@york.ac.uk

How to use this book

There are a number of features in the student books that we hope will help your learning. Some features will also help you to find your way around.

This A2 book covers the four A2 topics. These are shown in the contents list, which also shows you the page numbers for the main sections within each topic. There is an index at the back of the book to help you to find what you are looking for.

Main text

Key terms in the text are shown in **bold type**. These terms are defined in the interactive glossary on the software and can be found using the 'search glossary' feature.

There is an introduction at the start of each topic which provides a guide to the sort of things you will be studying in the topic.

There is an '**Overview**' box on the first spread of each topic, so you know which biological principles will be covered.

> **Overview** of the biological principles covered in this topic
>
> Understanding what is meant by biodiversity and species provides the starting point for this topic. You will look at the use of the binomial system, biological keys and classification for making sense of biodiversity.

Occasionally in the topics there are also '**Key biological principle**' boxes where a fundamental biological principle is highlighted.

> **Key biological principle:** Classification is dynamic
>
> Originally there were only two kingdoms, and then three: plants, animals and fungi. The separation of living things into five main groups, the five-kingdom classification, has been the generally accepted system since it was proposed by R. H. Whittaker in papers written in 1959 and 1969 and later modified by L. Margulis and K. V. Schwartz in 1982. No classification system will ever be perfect because new organisms are continuously being discovered, and because some organisms do not quite fit into particular groups, which causes controversy and disagreement.

'**Did you know?**' boxes contain material that will not be examined, but we hope you will find interesting.

> **Did you know?** How we can measure genetic variation
>
> Suppose you are running a zoo. One of the problems you face is ensuring that you conserve the genetic variation within the species in your zoo. To do this, you need to be able to measure genetic variation.
>
> There are two common measures of genetic variation. The first is the percentage of loci in a population with two or more alleles. The second is the percentage of loci that are heterozygous in the individuals of a population.

Questions

You will find two types of question in this book.

In-text questions occur now and again in the text. They are intended to help you to think carefully about what you have read and to aid your understanding. You can self-check using the answers provided at the back of the book.

> **Q5.11** In a species of plant, the allele for white flowers (**W**) is (unusually) dominant to the allele for pink flowers (**w**), and the allele for large leaves (**L**) is dominant to the allele for small leaves (**l**). Construct a genetic diagram to show the genotypes and phenotypes you would expect from a cross between a pure-breeding plant with large leaves and pink flowers and a heterozygous plant. As always when doing such crosses, take care that you distinguish your lower case and upper case (capital) letters!

> **Checkpoint**
>
> **5.2** Produce a concept map or table which summarises how genetic diversity is generated.

Boxes containing '**Checkpoint**' questions are found throughout the book. They give you summary-style tasks that build up some revision notes as you go through the student book.

Links to the software

Boxes that have an arrow icon link to features on the software. They are found in the margin near to the relevant piece of text.

Activity

In **Activity 5.6** you can sort out the features of the five kingdoms. **A5.06S**

'**Activity**' boxes show you which activities are associated with particular sections of the book. Activity sheets and any related animations can be accessed from an activity homepage for each activity, found via 'topic resources' on the software. There may also be weblinks to useful websites. Activity sheets include such things as practicals, issues for debate and role plays. They can be printed out. Your teacher or lecturer will guide you on which activity to do and when.

A final activity for each topic enables you to **check your notes** using the topic summary provided within the activity. The summary shows you what you need to have learned from each topic for your unit exam.

Weblink

You can find out more about the hotspots by visiting the Conservation International Hotspots website and using their explorer map.

'**Weblink**' boxes give you useful websites to go and look at. They are provided on a dedicated 'weblinks' page on the software which is found under 'SNAB communications'. You can also access them through the Heinemann website at www.heinemann.co.uk/hotlinks.

Extension

Extension 5.1 allows you to compare different ways of measuring biodiversity and calculate Simpson's indices for contrasting communities. **X5.01S**

'**Extension**' boxes refer you to extra information, associated with particular sections, which you can find on the software. The extension sheets can be printed out. The material in them will not be examined.

Support

To find out more about energy transfer in chemical reactions look at the Biochemistry support on the website.

'**Support**' boxes are provided now and again, where it is particularly useful for you to go to the student support provision on the software, e.g. biochemistry support. You will also be guided to the support on the software from the activity homepages, or you can go there directly via 'skills support'.

Topic test

Now that you have finished Topic 5, complete the end-of-topic test before starting Topic 6.

At the end of each topic, as well as the **check your notes** activity for consolidation of the topic, there is a '**Topic test**' box for an interactive topic test. This test will usually be set by your teacher/lecturer, and will help you to find out how much you have learned from the topic.

(5) On the wild side

Why a topic called On the wild side?

Much of an Advanced Level biology course concentrates on people. But people are only one part of a tremendously rich **biodiversity**. Today's biologists collect (Figure 5.1), quantify and classify the living world. Yet despite their best efforts and recent advances in scientific understanding we still have only a rather vague idea of how many millions of species share the planet – and even less idea of how species interact with one another.

Biodiversity in crisis

It has been estimated that there may be over 30 million different species but there is growing concern about loss of biodiversity, and with it the biological resources that we rely on. Even some species that were once very common are threatened. Fish and chips may be a British national dish but for how much longer? Many species, including cod and haddock, have declined to critical levels due to such human activities as habitat destruction and overexploitation. It has been estimated that the current **extinction** rate is between 1000 and 10 000 times higher than is naturally the case.

Should we be worried about this decline? How should we conserve the remaining species and the genetic diversity they contain? Putting up a fence around the remaining habitat and leaving well alone is rarely the answer, though it can help sometimes (Figure 5.2). We need knowledge of habitats, the species they contain, the environmental factors that influence them, the

Figure 5.1 A biologist dangles from ropes to find the Mallorcan midwife toad (*Alytes muletensis*). Once thought to be extinct, this toad is still vulnerable. Several hundred tadpoles and toads have been reintroduced to the wild.

Figure 5.2 Threatened Pacific tree snails (*Partula* spp.) are eaten by the introduced snail *Euglandina rosea*. Captive-bred *Partula* snails have been reintroduced to the wild. Small electric fences have been used to keep out the predatory *Euglandina* snails.

ways humans use and abuse them, and how they change over time. This helps us to understand how ecosystems work, and provides an insight into how to manage habitats for sustainable use of resources.

Why so many different organisms?

Two days after Christmas 1831, HMS *Beagle* eased away from her moorings in Plymouth. On board was a young man whose fascination with the natural world drove him to join the five-year voyage. His name was Charles Darwin (Figure 5.3) and he was to formulate a theory to explain a fundamental question in biology – why are there so many different kinds of organisms out there?

The voyage of the *Beagle* took Darwin to many countries. He discovered numerous new species, many of them in the Galapagos Islands, lying 1000 km off the west coast of South America. Darwin found many new species on the islands – species that were similar to but not the same as those on the South American mainland. As he examined these species he began to question what he saw. Over the next 20 years he refined a theory that would explain his observations, eventually publishing his ideas in his book *The Origin of Species*.

▲ **Figure 5.3** Charles Darwin in his early twenties.

◄ **Figure 5.4** How would you classify this animal and why? Complete Activity 5.2 to check your ideas.

Overview of the biological principles covered in this topic

Understanding what is meant by biodiversity and species provides the starting point for this topic. You will look at the use of the binomial system, biological keys and classification for making sense of biodiversity.

Genetic diversity, inheritance, natural selection and evolution provide a mechanism for the formation of a wealth of biodiversity. Knowledge of photosynthesis and the transfer of energy through ecosystems provides a basis for understanding how ecosystems work.

You will investigate the ecological factors that influence which species occur in a particular habitat, and study how species are adapted to cope with the conditions. You will look in detail at succession and how the changes it produces affect biodiversity. The role of zoos in the conservation of endangered species both in captivity and in the wild will be examined.

Activities

In **Activity 5.1** take a tour of the Galapagos Islands to see the variety of life that Darwin encountered. **A5.01S**

Activity 5.2 lets you develop some of the observation and interpretation skills used by both Darwin and today's biologists. **A5.02S**

5.1 What is meant by biodiversity?

Biodiversity is used in everyday language to mean the variety of life, and in particular the wealth of different **species** that exist. However, in biological terms it refers to:

- the variety of different organisms
- the diversity within species (**genetic diversity**)
- the diversity of **ecosystems**.

The 1992 International Convention on Biological Diversity defined biodiversity as:

> *the variability among living organisms from all sources including, inter alia, terrestrial, marine and other aquatic ecosystems and the ecological complexes of which they are part; this includes diversity within species, between species and of ecosystems.*

(Note: *inter alia* is Latin for 'among other things'.)

One of the more surprising facts about life on Earth in the twenty-first century is that we still have no rigorous estimate of how many species we share the planet with. This is not because of any particular technical difficulties. It's simply because the size of the task is massive. Somewhere between 1.4 and 1.7 million species have already been described and named (Figure 5.5).

Activities

In **Activity 5.3** you can consider different ways of defining biodiversity. **A5.03S**

Activity 5.4 lets you read about biologists involved in biodiversity research and helps you investigate the rate at which new species are discovered. **A5.04S**

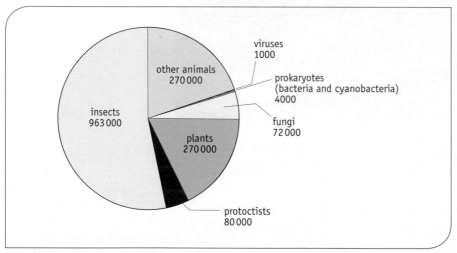

🔺 **Figure 5.5** Estimates of the numbers of species currently known show insects dominating, although it is thought that there are large numbers yet to be discovered, particularly of bacteria and fungi.

Current estimates of the actual number of species on Earth range from 5 million to 30 million or more! In just one animal group alone, the beetles, there are thought to be at least 1 million and perhaps as many as 5 million species. So far, only about 450 000 beetle species have been described and named, less than half the world's beetle diversity. The situation may be far worse in other less well studied or less popular groups such as bacteria and fungi, where probably only a tiny fraction of the species have been given a name.

Producing a species catalogue

The international Species 2000 programme was set up in September 1994. Its objective was to make a catalogue of all the known species of plants, animals, fungi and microbes on Earth as the baseline dataset for studies of global biodiversity. By 2000 its checklist contained 220 000 species. This had risen to 304 000 by 2003, which is a sizeable proportion of the estimated 1.4–1.7 million species that have been described and named. The programme was established by the International Union of Biological Sciences (IUBS) and is based in the School of Plant Sciences, University of Reading.

The efforts of the Species 2000 programme may seem impressive compared with the mere 12 000 or so species described by Carolus Linnaeus. Linnaeus, a great Swedish biologist, made the first attempt to produce such a global species catalogue (*Systema Naturae*, published in 1758). However, the task still facing the Species 2000 programme remains huge with as many as 30 million species still to be described. This means that a great deal more exploration, discovery and naming has still to be done by taxonomists.

New technology will help speed up the process. Geographical positioning systems (GPSs) can quickly pinpoint the location of new species whilst the Internet can be used to store and exchange data from around the world. Crucially, DNA electrophoresis can now be used to see if 'new' species are indeed different from existing species.

Key biological principle: What is a species?

Figure 5.6 This is not a mouse! Although it looks like one, it is a marsupial and is more closely related to kangaroos and koalas than to rodents.

What is a species? Perhaps surprisingly, this is a difficult question for biologists to answer. There are a variety of definitions used. As you can see from Figure 5.6, it is not enough to say that a species is a group of individuals that closely resemble one another. The most widely used definition is:

A species is a group of organisms, with similar morphology, physiology and behaviour, which can interbreed to produce fertile offspring, and are reproductively isolated (in place, time or behaviour) from other species.

(Continued)

Although horses and donkeys have, to some extent, similar morphologies (features), physiologies and behaviours, and can be bred together to produce mules, the mules are infertile. Therefore horses are closely related to donkeys but remain distinct species. However, some species do not seem to know this rule and will quite happily breed and produce hybrid offspring that are fertile. Even Darwin's famous finch species of the Galapagos Islands have been known to interbreed and produce viable offspring. If this happened frequently we would expect small islands like Daphne, with only four species of finch, to end up with birds of average types and no distinct species. However, the finch species remain distinct because the hybrid forms do not survive drought conditions as well as the pure-bred types. There is enough fluctuation in the island's climate to maintain the different species.

Biologists sometimes disagree about a particular species and sometimes one species is reclassified into two. For example, it has recently been realised that the UK has not one but two species of pipistrelle bat. These can most easily be distinguished by the frequency of their calls. One species (*Pipistrellus pipistrellus*) is known as '45 kHz' and the other (*P. pygmaeus*) as '55 kHz'. Marsh and willow tits were also once thought to be the same species (Figure 5.7).

🔺 **Figure 5.7** The willow tit (left, *Parus montanus*) was thought to be a marsh tit (right, *Parus palustris*) but they are now classified as two separate species.

Sometimes two separate species are reclassified as a single species. For example, the locust species *Locusta migratoria* and *Pachytylus migratoria* were merged into one species, known as the migratory locust (*Locusta migratoria*).

Q5.1 Read the descriptions below and decide whether the groups are the same species or different species. Explain your answers.
a Group A has 98% of its genes in common with group B.
b Group C can interbreed with group A and produce fertile offspring.
c Group D is physically very similar to group B except that it is a different colour. Where groups B and D overlap in geographical areas, matings between them produce offspring of an intermediate colour. The males of these offspring are infertile.
d Group E looks very similar to group F but has darker feathers. The genetic fingerprints of groups E and F are identical.

Did you know? Where do estimates of species numbers come from?

Estimates of the possible total numbers of species are not simply guesswork, but can be described as informed 'guestimates'. A number of different approaches have been used and all have limitations. One approach is to generate global data by extrapolating from figures for a well studied region. For example, in the UK, one of the best catalogued countries in the world, 22 000 species of insect have been described, including 67 butterfly species, giving a UK butterfly-to-all-insects ratio of 1 to 328. Because butterflies are very conspicuous insects, they are well catalogued around the world, and about 17 500 have been named. If the global ratio of butterfly species to all insect species is the same as the UK ratio, then the global insect species total should be around 17 500 × 328 or 5.74 million. However, this method will only give a reliable estimate if the global ratio is similar to the UK ratio.

A different approach is to multiply up from an intensive survey of a very small sample area. A survey of beetle species in the tropical forests of Panama found that 160 species were only ever found in one particular species of tree. Since beetles make up about 40% of all insect species, there could be up to 400 different insect species that are unique to the canopy of this single tree species. If this is true for every one of the 50 000 or so species of tropical forest trees, there could be up to 20 million insect species in the global tropical forest canopy alone.

These two estimates for global insect biodiversity differ greatly and suggest that most global biodiversity estimates must be viewed cautiously.

Q5.2 Suggest why these two methods of estimating species number may not give accurate values for total species numbers.

5.2 Organism diversity

Examining biodiversity often means simply considering the variety of different species present in a given area. To do this biologists need to be able to identify, name and classify the species they observe.

Unlocking identities: unique names

Putting a name to the species they find represents a significant challenge to biologists working in the field. For any biologist, from a bird-watcher looking at wildfowl on an estuary in the UK to a research worker studying ants in a tropical rainforest, accurate biological identification is vital.

All organisms are given a scientific name. This avoids the confusion that can arise when common names are used. For example, *Arum maculatum* (Figure 5.8) has more common names than any other plant in Britain, including lords and ladies and cuckoo pint. Probably the most important thing that Carolus Linnaeus, the Swedish biologist, did in his early attempts to catalogue living things was to come up with a system in which each species was given a unique two-part Latin name. This **binomial system** is still in use today. The first part of the name, the **genus**, is shared by all closely related species, so all horses and zebras are in the genus called *Equus*. The second part of the name defines the particular species in the genus. Together these two words make up a unique species name.

Scientific names are often highly descriptive. For example, the greater horseshoe bat in Figure 5.9 is called *Rhinolophus ferrumequinum*, which perfectly describes its fleshy horseshoe-shaped nose! So the name is not only a unique scientific label for that species, that can be understood in any part of the world, but also contains an apt description of the bat's distinguishing feature. This bat will have different common names in different regions, but use of its scientific name prevents any possible confusion.

▲ **Figure 5.8** More than a dozen common names have been recorded for *Arum maculatum*. Species names are always written in italics or underlined. The first part of the name, the genus, always starts with a capital letter. The second part of the name does not.

◀ **Figure 5.9** This bat has the scientific name *Rhinolophus ferrumequinum*, from *rhinos* (Greek for nose), *lophos* (Greek for crest), *ferrum* (Latin for iron) and *equinum* (Latin for horse) providing an excellent description of its nose – if you understand Greek and Latin.

Q5.3 Using Table 5.1:
a state to which genus creeping buttercup belongs
b decide whether meadowsweet and dropwort or lady's mantle and dropwort are more closely related, and give a reason for your answer.

Table 5.1 Common and scientific names for some British wild flowers.

Common name	Scientific name
creeping buttercup	*Ranunculus repens*
meadowsweet	*Filipendula ulmaria*
lady's mantle	*Alchemilla vulgaris*
dropwort	*Filipendula vulgaris*

Biological identification

To name an organism you have to be able to identify what it is. Keys are a traditional means of identifying organisms. You will no doubt be familiar with the use of a simple text-based **dichotomous key**, which gives two alternatives at each stage in the key. Versions of this type of key are routinely used by field biologists to find the name of unfamiliar organisms. An example is shown in Figure 5.10.

Q5.4 Use the key in Figure 5.10 to identify the butterfly fish shown in Figure 5.11.

Traditional dichotomous keys have to be used in a set sequence, starting with the first pair of statements. Sometimes this can be a problem because if a character referred to early in the key is absent, due to damage or the stage of development of the organism, it is difficult to continue to use the key. Multiple access keys are more flexible. They allow the user to start with whichever character they wish and work through the key in a different order.

Often regarded as one of the most traditional areas of biology, identification methods are increasingly making use of new technology. The field of computer-assisted taxonomy (CAT), which uses computer-based keys and electronic techniques, is expanding and beginning to replace paper-based keys. Computer-based keys can allow multiple access and easily be updated when new species are discovered.

Activity

Use the SAPS computer-based key in **Activity 5.5**. A5.05S

Sorting and grouping

A hierarchical system

Faced with such a bewildering variety of living things, humans have always tried to organise and make sense of the variety of life. Placing organisms into groups based on shared features, known as classification or **taxonomy**, results in a more manageable number of categories and has been the principal aim of all classification systems. However, it was Linnaeus who created the hierarchical system in use today.

On coral reefs there is often an amazing array of colourful fish. But on closer inspection, similarities between groups of fish become apparent. Look at the butterfly fish in Figure 5.12. What features do they all have in common? They are thin-bodied fish, many having a dark band across the eye. Each has a small mouth at the end of an extended snout, and tiny bristle-like teeth. The three butterfly fish in Figure 5.12 all have these features and all belong to the genus *Chaetodon*. Look at Figure 5.12 and decide if you can see the similarities.

Decision point	Choices	Decision
1	Pelvic fin dark Pelvic fin light	2 4
2	Two large white spots below dorsal fin Lacks two large white spots below dorsal fin	*C. quadrimaculatus* 3
3	Tail with two dark bars at tip Tail with one dark bar at tip	*C. reticulatus* *C. kleinii*
4	Posterior or dorsal fin has long filament extension Filament extension lacking from dorsal fin	5 6
5	Large dark spot on body near filament Small dark spot on body near filament	*C. ephippium* *C. auriga*
6	No vertical band through eye Vertical band through eye	*C. fremblii* 7
7	Incomplete eyeband on face (does not go to top of head) Complete eyeband on face (extends to top of head)	*C. multicinctus* 8
8	Nose area with band Nose area lacks band	9 10
9	Fewer than eight diagonal bands on body More than eight diagonal bands on body	*C. ornatissimus* *C. trifasciatus*
10	Distinct white spot splits eyeband above eye No white spot above eye; eyeband not split	*C. lineolatus* 11
11	Upper third of body under dorsal fin dark Upper third of body under dorsal fin not dark	*C. tinkeri* 12
12	Distinct small spots arranged in rows No distinct small spots; body has large spot or band	13 14
13	No black band on caudal fin Obvious black band on caudal peduncle	*C. citrinellus* *C. miliaris*
14	Side with a large black teardrop; no dark bars on tail Large black shoulder patch; tail with dark bars	*C. unimaculatus* *C. lunula*

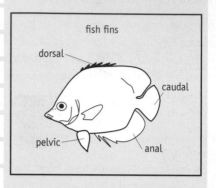

fish fins

dorsal

caudal

pelvic

anal

▲ **Figure 5.10** Dichotomous key for butterfly fish of the genus *Chaetodon*. *Source: University of Hawaii.*

◀ **Figure 5.11** Use the key above to identify this butterfly fish species.

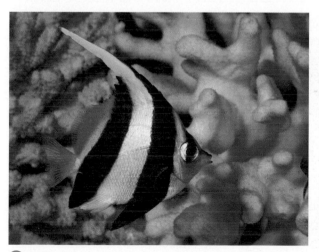

🔺 **Figure 5.12 A** The archer butterfly fish (*Chaetodon bennetti*), **B** the white spotted butterfly fish (*Chaetodon kleinii*) and **C** the limespot butterfly fish (*Chaetodon unimaculatus*). These butterfly fish share common characteristics and all belong to the same genus.

Q5.5 Look at Figure 5.13. What distinctive feature do the bannerfish in the genus *Heniochus* seem to have in common?

🔺 **Figure 5.13** Bannerfish are related to the butterfly fish in Figure 5.12 but are sufficiently different to be classified in a different genus.

The bannerfish shown in Figure 5.13 look rather like the butterfly fish in Figure 5.12. However, they are sufficiently different to be grouped in a different genus, *Heniochus*. They also share a number of similarities with members of the genus *Chaetodon*, like the butterfly fish, so the two genera are grouped together in the same **family** – the Chaetodontidae (butterfly fishes).

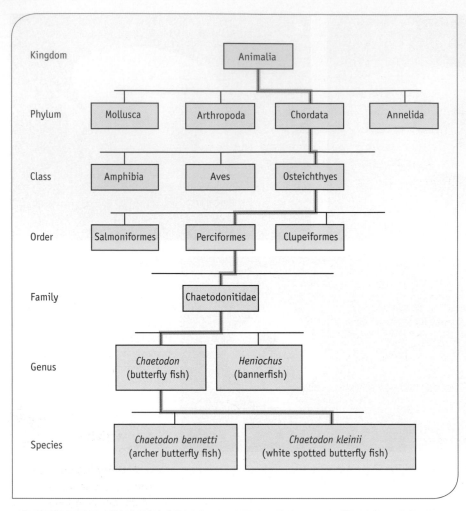

Figure 5.14 Butterfly fish classification from kingdom to species. (Don't worry, you don't need to remember the names!)

Together with many other reef fish, members of this family all have thoracic as opposed to abdominal pelvic fins (that is, their lower fins are towards the front of the body). All families sharing this feature are grouped in the **order** Perciformes (perch-like), the largest order of fish, and all the orders of fish with bony skeletons are grouped together in the **class** Osteichthyes (bony fish). This class, along with all the other classes of animals with a dorsal spinal cord, is placed in the **phylum** Chordata – see Figure 5.14. All animals are grouped in the **kingdom** Animalia.

So the taxonomic hierarchy is a series of nested groups or **taxa** (singular **taxon**), in which the members all share one or more common features or **homologies**. It is similar to the way in which you might arrange the folders and files in your work area of a computer (assuming you are that organised!).

The hierarchy of groups is:

- kingdom
- phylum (plural phyla)
- class
- order
- family
- genus (plural genera)
- species.

Five kingdoms and counting

Butterfly fish are clearly animals. Along with many other organisms, we readily recognise them as members of the kingdom **Animalia**. In the same way, we can easily recognise the sea grass growing in the shallow waters around a coral reef as a member of the kingdom **Plantae**. At this level, everyone uses the classification system automatically because we grow up with the idea of what plants and animals are. What many tourists gazing at the reef do not know is that there are generally held to be three more kingdoms, the **Fungi**, the **Protoctista** and the **Prokaryotae**. The five kingdoms used today, Figure 5.15, represent the top level in the taxonomic hierarchy.

Kingdom Animalia

- Multicellular eukaryotes with differentiated cells organised into specialised organs.
- No cell walls or large vacuoles.
- Cannot photosynthesise.
- Heterotrophic, relying on other organisms for nutrition.
- Most can move from place to place and have nervous coordination.
- Includes phyla such as jellyfish, roundworms, arthropods and molluscs.

Kingdom Plantae

- Multicellular eukaryotes with differentiated cells organised into specialised organs.
- Cell walls contain cellulose.
- Cells contain chloroplasts and large vacuoles.
- Autotrophic, making organic compounds by photosynthesis (except for a few parasites).
- Includes mosses, liverworts, ferns, conifers and flowering plants.

Kingdom Fungi

- Multicellular eukaryotes (although most do not have separate cells).
- Most are made up of a network of thread-like strands, called multinucleate hyphae.
- Cell walls made of chitin, a mucopolysaccharide (similar to cellulose but with amino acids attached).
- Cannot photosynthesise.
- Heterotrophic – most absorb nutrients from decaying matter after extracellular digestion.
- Includes moulds, yeasts and mushrooms.

Kingdom Protoctista

- Multicellular and unicellular eukaryotes.
- Basic body structure is relatively simple.
- May either photosynthesise or feed on organic matter from other sources.
- Includes single-celled protozoa, such as *Amoeba* and *Paramecium*, and algae.

Kingdom Prokaryotae

- Prokaryotic cells are very small, typically less than 10 micrometres across.
- Cells have no distinct nucleus. The nucleic acid is in a single circular chromosome.
- Cells do not have organelles such as mitochondria or chloroplasts.
- May either photosynthesise or feed on organic matter from other sources.
- Includes the bacteria and blue-green bacteria (Cyanobacteria).

 Figure 5.15 The five kingdoms and their key features.

Activity

In **Activity 5.6** you can sort out the features of the five kingdoms. **A5.06S**

Activity

When Darwin visited the Galapagos Islands, he met many unusual and unfamiliar organisms. Initially he had to use just their visible features and any evidence he could gather from their surrounding environment and lifestyle to decide how to classify them. **Activity 5.7** sees if you can assign some of Darwin's discoveries to the correct taxonomic group. **A5.07S**

Originally there were only two kingdoms, and then three: plants, animals and fungi. The separation of living things into five main groups, the five-kingdom classification, has been the generally accepted system since it was proposed by R. H. Whittaker in papers written in 1959 and 1969 and later modified by L. Margulis and K. V. Schwartz in 1982. No classification system will ever be perfect because new organisms are continuously being discovered, and because some organisms do not quite fit into particular groups, which causes controversy and disagreement.

The separation of the single-celled eukaryotic organisms and their relatives into their own kingdom, the Protoctista, and the grouping of bacteria into a fifth kingdom of their own, the Prokaryotae, is the result of trying to achieve a system which reflects the real distinctions between organisms, especially at the cellular level. As we saw in Topic 3, the Prokaryotae were recognised to have a completely different cellular organisation from eukaryotic organisms, so they were placed into their own kingdom.

More recent discoveries about the molecular biology of bacteria have resulted in the suggestion that there should be a sixth kingdom. Some bacteria, the heat-loving sulphur-users and methane-producers, have sufficient differences in their nucleic acids for them to be regarded as distinct from all other organisms. Most of these sorts of bacteria are anaerobic and are thought to be surviving members of the oldest kinds of living organism on Earth, from a period when there was no oxygen in the atmosphere. The suggestion is that these organisms should be placed in a kingdom of their own, the **Archaebacteria**. This would mean that three large groups of organisms would be recognised – Archaebacteria, Prokaryotae and Eukaryotae – rather than six kingdoms. So although it may take a while for this to be generally accepted, the classification system is again changing to take account of our new understanding.

Why classify?

As well as being a logical way to organise biodiversity, such as the shoals of brightly coloured fish around a coral reef, classification groups have a deeper biological significance. Since members of a taxon share common features, the great majority of biologists accept that they also share a common evolutionary ancestor (the exceptions being those biologists who do not accept the theory of **evolution**). This means that, in evolutionary terms, members of a taxon are more closely related to each other than they are to any organisms outside that taxon. In terms of the butterfly fish, members of the genus *Chaetodon* are all more closely related to each other than to members of the genus *Heniochus*, and all of these fish are more closely related to one another than they are to members of other fish families. Diagrams of classification tell an evolutionary story, rather like a family tree.

Identifying closely related species can be very valuable, for example when trying to find new sources of chemicals with medicinal or other beneficial properties. An antiviral drug was identified in an Australian tree species, the Moreton Bay chestnut, but was found to be toxic. However, plants in the closely related genus *Alexa* were found to contain the drug in a less toxic form.

Checkpoint

5.1 Produce a list of the principles of taxonomy.

Taxonomy and measuring biodiversity

When considering biodiversity it is usual simply to look at the variety of different species that exist in a given area, so that an area with a greater number of species is said to have a higher biodiversity. However, consider the hypothetical situation where a biologist comparing two sites finds the species shown in Table 5.2.

▼ **Table 5.2** Number of species at two sites.

Number of species	Site 1	Site 2
plants	80	50
beetles	20	0
insects	0	20
birds	0	15
spiders	0	5
molluscs	0	5
mammals	0	5
total number	100	100

Both sites contain 100 species but the second surely has a greater biodiversity. In considering diversity of organisms biologists may not look just at the number of different species, but may also consider the numbers of individuals, populations or higher taxonomic groups – genera, families, phyla, and so on. They may also look at the genetic diversity.

Did you know? How we quantify biodiversity

Assessing biodiversity
However biologists account for the great variety of life we find before us, there are many different ways of assessing biodiversity. The simplest and most commonly used approach is to count the number of species present. This provides a measure of species richness.

Many other methods involve the calculation of some sort of diversity index. One example is Simpson's index, which is based on the probability that a second organism collected from a community will be in the same species as the first. Alternative approaches consider the genetic diversity of an area by counting the number of different groups of organism present, or score the rarity of the species present by looking at how patchily or how evenly they are distributed.

Once biodiversity has been assessed, the data can be mapped on both a local and a global scale to reveal patterns in diversity. This can help to focus conservation efforts on vulnerable habitats or species.

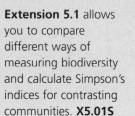

Extension

Extension 5.1 allows you to compare different ways of measuring biodiversity and calculate Simpson's indices for contrasting communities. **X5.01S**

Finding the hotspots
The world's biodiversity is not distributed evenly across the surface of the planet (Figure 5.16). In the last few decades, discoveries have revealed unexpectedly high levels of biodiversity in unlikely places. For example, while coral reefs and rainforests are ranked as the richest ecosystems in the world, the **biodiversity hotspot** containing the largest proportion of the world's plants is not, as you might expect, in the tropics. Instead, the plant hotspot is the Mediterranean Basin which contains one in ten of the Earth's plant species. More than half are **endemic**, which means they are found only in that area and nowhere else on Earth. Another example comes from ocean sea beds. These have long been regarded as very low-diversity ecosystems. However, exploration of the muds of deep oceans has revealed a startling range of animals living in sediments previously thought to be lifeless.

(Continued)

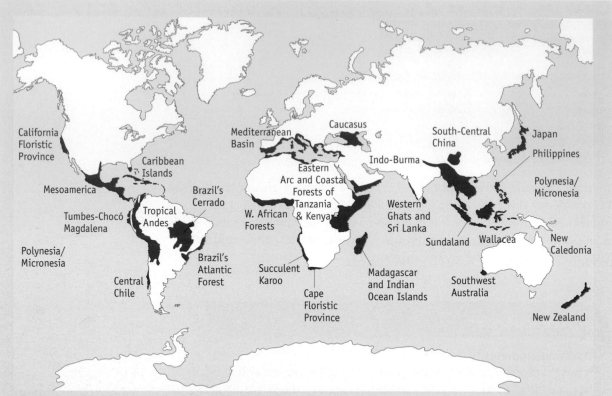

△ **Figure 5.16** Biodiversity hotspots of the world. *Source: Nature.*

The term 'biodiversity hotspot' was coined in 1988 by conservationist Norman Myers to describe areas of particularly high biodiversity. He identified ten tropical forest regions containing between them 13% of all plant diversity in just 0.2% of the Earth's land area. Conservation International later adopted the idea as a way of focusing conservation effort on the most critical places. They have extended the list to include 34 regions collectively covering only about 2% of the Earth's land surface, yet containing more than 50% of all terrestrial species. These particular hotspots are identified in terms of plant diversity. To be regarded as a hotspot according to these criteria, a region must have a minimum of 1500 different plant species; crucially, it must also have 0.5% or more of total global plant diversity present as endemic species.

Q5.6 Why are hotspots identified only in terms of plant diversity?

Q5.7 **a** Describe a plausible explanation for the distribution of hotspots shown on the map in Figure 5.16.
b As you will see from the map, many of the hotspots are over islands. Why do you think islands contain a relatively higher level of species endemism than equivalent areas of land on the main continental landmasses?

Weblink

You can find out more about the hotspots by visiting the Conservation International Hotspots website and using their explorer map.

5.3 What do we mean by genetic diversity?

Species are made up of individuals that differ from one another – they show variation. Topic 3 discussed how the appearance of an organism, its **phenotype**, is the result of interactions between its **genotype** and the **environment**. In all organisms that reproduce sexually, every individual (except for cases such as identical twins and cloned organisms) has a different combination of alleles. This is known as **genetic diversity**; the greater the variety of genotypes the more genetically diverse the population.

Sources of genetic variation

Looking at species in their natural habitats, such as zebras, lions and acacia trees in the savannah, it is easy to overlook genetic differences within each species because the differences *between* species are more noticeable (Figure 5.17). Yet you only need look around at the other members of your advanced biology group to realise just how much variation there can be ...

Much genetic variation has no measurable effect on phenotype. It is made up of molecular differences which can only be detected using techniques such as gel electrophoresis of proteins, with the two alleles at a single locus producing slightly different protein products. However, these differences may be important in evolutionary terms.

Where does all this genetic diversity come from? Figure 5.18 outlines sources of variation.

 Figure 5.17 At first glance these zebras may all look the same but on closer inspection there are clear differences between the individuals. Some of this variation may be genetic in origin.

Mutations

Mutations change the DNA sequence in the cells of an organism and create new alleles. Genetic diversity is increased with the addition of these new alleles to the **gene pool** – which consists of all the alleles of all the genes present in a population. In Topic 2 we saw how the deletion of three nucleotides was one of the mutations causing cystic fibrosis. We also saw how a single point mutation (alteration of one base) could result in the formation of malfunctioning haemoglobin, giving rise to sickle cell anaemia.

Most mutations have harmful effects. However, some mutations can be beneficial to the organism. For example, mutations in houseflies which make them resistant to the pesticide DDT are an advantage to these individuals when DDT is present in the environment. Even the sickle cell anaemia mutation can be advantageous. The heterozygotes who carry both sickle cell and normal alleles are more resistant to malaria. In countries where malaria is widespread, carrying the sickle cell allele gives a distinct advantage.

Extension

Approximately one in seven human mutations are caused by jumping genes. To find out more read **Extension 5.2**. **X5.02S**

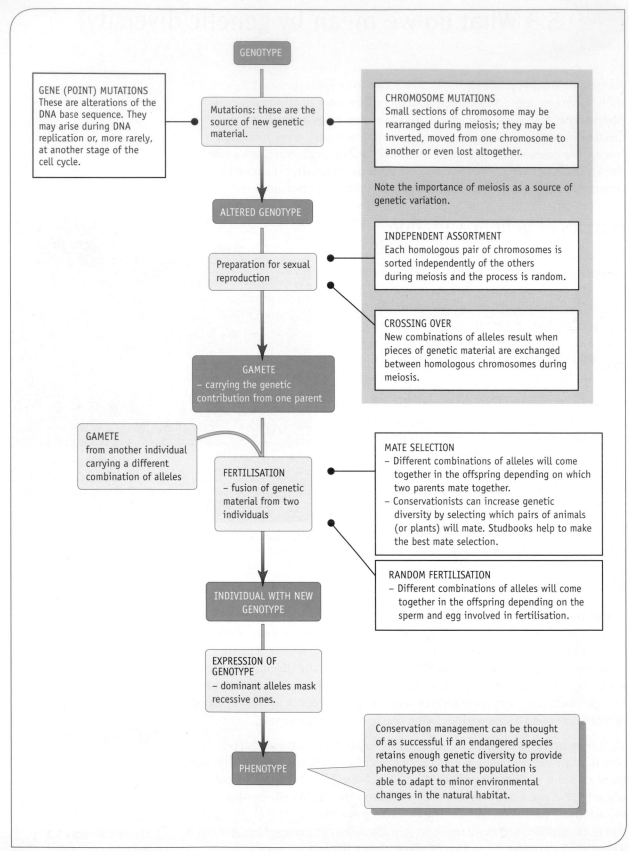

Figure 5.18 Sources of genetic variation. The random processes result in greater genetic variation in a population.

Recombination of genes

The shuffling of existing genetic material into new combinations is important in creating genetic variation. This shuffling takes place in the preparations for sexual reproduction, and involves rearrangement of the genetic material during meiosis. This includes:

- **independent assortment** and
- **crossing over**.

Independent assortment

As we saw in Topic 3, during the process of meiosis only one chromosome from each homologous pair ends up in each gamete. This process is random; either chromosome from each pair could be in any gamete. This way of sharing out chromosomes produces genetically variable gametes.

Crossing over

During the first meiotic division, homologous chromosomes come together as pairs and the **chromatids** come into contact. At these contact points the chromatids break and rejoin, exchanging sections of DNA (Figure 5.19). The point where the chromatids break are called **chiasma** (plural **chiasmata**), and several of these often occur along the length of each pair of chromosomes, giving rise to a large amount of variation.

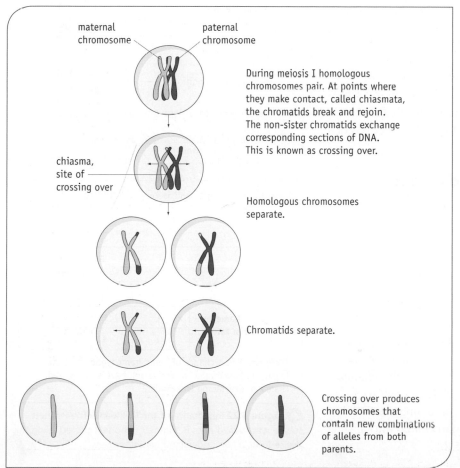

maternal chromosome

paternal chromosome

During meiosis I homologous chromosomes pair. At points where they make contact, called chiasmata, the chromatids break and rejoin. The non-sister chromatids exchange corresponding sections of DNA. This is known as crossing over.

chiasma, site of crossing over

Homologous chromosomes separate.

Chromatids separate.

Crossing over produces chromosomes that contain new combinations of alleles from both parents.

Figure 5.19 Crossing over can result in a great deal of genetic variation.

Q5.8 Look at Figure 5.20. What would the chromosomes in the gametes produced from the homologous chromosomes in cells **A** and **B** look like after crossing over has occurred?

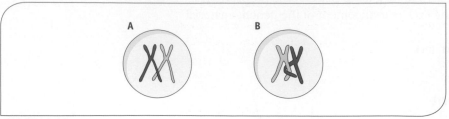

🔺 **Figure 5.20** Crossing over.

Inheritance and genetic variation

The human genotype probably contains between 20 000 and 25 000 genes. We shall illustrate the way these can be inherited and generate variation by considering the inheritance of just two genes – **dihybrid inheritance**. Before doing this, you might find it useful to study the basic 'genetic vocabulary' outlined in Figure 5.21.

The principle for understanding dihybrid crosses is that *the inheritance of one pair of alleles does not affect the inheritance of the other pair* because chromosomes act independently of one another during gamete formation. In other words, we are only considering genes carried on separate chromosomes – or sufficiently far apart on the same chromosome that crossing over means that the two genes effectively behave as if they were on different chromosomes.

Reminder: monohybrid crosses

The recessive form of the leuk gene contributes in humans to a tendency to develop leukaemia. Consider a cross between two **heterozygotes** (genotype **Ll**). During meiosis, the chromosomes in a pair separate, carrying the alleles to different gametes. The heterozygous female's eggs have either the **L** allele or the **l** allele. The same applies to the sperm cells produced by the heterozygous male. This separation of the alleles of one gene into different gametes during meiosis is known as the segregation of alleles. For each parent, the chance of a gamete containing the **L** allele is a half (0.5), as is the chance of a gamete containing the **l** allele. These probabilities can be incorporated into a Punnett square – see Figure 5.22.

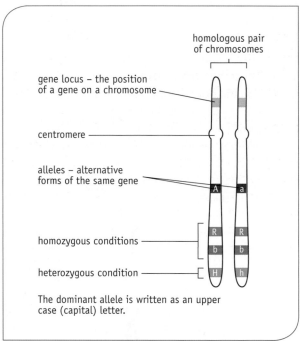

🔺 **Figure 5.21** Some basic genetic vocabulary.

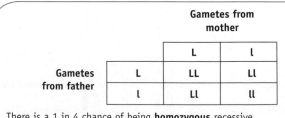

		Gametes from mother	
		L	**l**
Gametes from father	**L**	LL	Ll
	l	Ll	ll

There is a 1 in 4 chance of being **homozygous** recessive. If you are homozygous recessive you have an increased chance of developing leukaemia.

🔺 **Figure 5.22** Inheritance of the leuk gene. The Punnett square was named after R. Punnett, who first started using this device to help him understand genetic crosses.

Dihybrid crosses

A dihybrid cross can be considered as two monohybrid crosses occurring at the same time. In guinea pigs, one gene codes for hair colour. There are two alleles: **B** for black hair and **b** for brown hair. **B** is dominant and **b** is recessive. A second gene at a different **locus** (place) on a different chromosome codes for hair length; it also has two alleles, **S** for short hair and **s** for long hair. **S** is dominant and **s** is recessive.

Crossing two homozygotes

If a homozygous black short-haired guinea pig is crossed with a homozygous brown long-haired guinea pig, all the offspring will be heterozygous, **BbSs**, with black short hair, as shown below.

	Male	**Female**
Parental phenotypes	black, short-haired	brown, long-haired
Parental genotypes	**BBSS**	**bbss**
Gamete genotypes	all **BS**	all **bs**
Offspring genotypes	all **BbSs**	
Offspring phenotypes	all black, short-haired	

Crossing a heterozygote and a homozygote

When a heterozygote undergoes meiosis to form gametes, the two pairs of chromosomes line up on the equator independently of each other. There are two possible ways that the pairs can line up on the equator (Figure 5.23). Which way round they line up is completely random. With large numbers of cells undergoing meiosis, about half will line up one way and the other half the other way. This results in equal quantities of four different gametes being produced, as shown below.

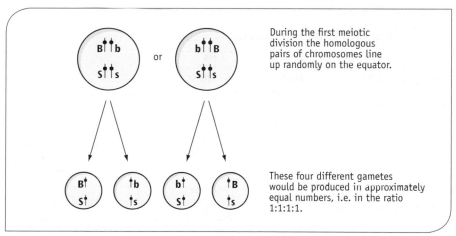

During the first meiotic division the homologous pairs of chromosomes line up randomly on the equator.

These four different gametes would be produced in approximately equal numbers, i.e. in the ratio 1:1:1:1.

Figure 5.23
Independent assortment leads to a variety of different genotypes in the gametes.

At fertilisation each male gamete has an equal chance of fusing with each female gamete.

If a heterozygous male, **BbSs**, was crossed with a homozygous recessive female, **bbss**, how many different genotypes and phenotypes would be created? The heterozygote produces four different gametes; the homozygote only one. So four offspring genotypes are produced with an average ratio of 1:1:1:1 or a 0.25 probability of each occurring, as shown overleaf.

	Male	Female
Parental phenotypes	black, short-haired	brown, long-haired
Parental genotypes	**BbSs**	**bbss**
Gamete genotypes	**BS Bs bS bs**	all **bs**
	equal proportions of each	

Possible genotypes after fertilisation:

Gametes from female

		bs
	BS	**BbSs**
Gametes	**Bs**	**Bbss**
from male	**bS**	**bbSs**
	bs	**bbss**

Q5.9 What are the phenotypes of the offspring in the above cross, and in what ratio do they occur?

Crossing two heterozygotes

When two heterozygous individuals are crossed the phenotypes and genotypes formed are as shown below.

	Male	Female
Parental phenotypes	black, short-haired	black, short-haired
Parental genotypes	**BbSs**	**BbSs**
Gamete genotypes	**BS Bs bS bs**	**BS Bs bS bs**
	equal proportions of each	equal proportions of each

Possible genotypes and phenotypes after fertilisation:

Gametes from one parent

		BS	**Bs**	**bS**	**bs**
Gametes from other parent	**BS**	**BBSS** black, short	**BBSs** black, short	**BbSS** black, short	**BbSs** black, short
	Bs	**BBSs** black, short	**BBss** black, long	**BbSs** black, short	**Bbss** black, long
	bS	**BbSS** black, short	**BbSs** black, short	**bbSS** brown, short	**bbSs** brown, short
	bs	**BbSs** black, short	**Bbss** black, long	**bbSs** brown, short	**bbss** brown, long

There are four phenotypes, and they would be expected to occur in the following ratio:

- 9 black, short-haired
- 3 black, long-haired
- 3 brown, short-haired
- 1 brown, long-haired.

This is known as a 9:3:3:1 ratio. This is the typical ratio of phenotypes for a dihybrid cross between two heterozygous individuals.

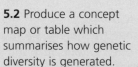

Checkpoint

5.2 Produce a concept map or table which summarises how genetic diversity is generated.

Now look at the genotypes of these offspring, and count up how many different genotypes occur. You should find that there are nine different combinations of the four alleles, that is, nine different genotypes. There is greater genetic diversity in the offspring of this cross, compared with any crosses involving homozygous individuals.

Activity

Try working out the outcomes for the dihybrid crosses in **Activity 5.8**. A5.08S

Q5.10   In a species of dog, the allele for curly tails (**C**) is dominant to the allele for straight tails (**c**), and the allele for large ears (**L**) is dominant to the allele for small ears (**l**). If a male dog, homozygous dominant at both loci, was mated with a female dog, heterozygous at both loci, what genotypes and phenotypes would you expect in the offspring?

Q5.11   In a species of plant, the allele for white flowers (**W**) is (unusually) dominant to the allele for pink flowers (**w**), and the allele for large leaves (**L**) is dominant to the allele for small leaves (**l**). Construct a genetic diagram to show the genotypes and phenotypes you would expect from a cross between a pure-breeding plant with large leaves and pink flowers and a heterozygous plant. As always when doing such crosses, take care that you distinguish your lower case and upper case (capital) letters!

Did you know? Multiple alleles and codominance

Some genes have more than two alleles, known as multiple alleles. A frequently described example is the gene for human blood groups. The four blood groups are controlled by a single gene which has three alleles: I^A, I^B and I^O. The inheritance of multiple alleles creates greater diversity.

This example also illustrates the phenomenon of codominance. In codominance, the phenotype of the heterozygote reflects the presence of both alleles. In the case of blood groups, individuals with the alleles $I^A I^B$ have blood group AB. I^A and I^B are codominant. Interestingly, though, both I^A and I^B are dominant to I^O.

Did you know? How we can measure genetic variation

Suppose you are running a zoo. One of the problems you face is ensuring that you conserve the genetic variation within the species in your zoo. To do this, you need to be able to measure genetic variation.

There are two common measures of genetic variation. The first is the percentage of loci in a population with two or more alleles. The second is the percentage of loci that are heterozygous in the individuals of a population.

The proportion of genes present in the heterozygous form can be expressed as a number called the heterozygosity index. This number can be calculated by examining DNA fingerprints (Figure 5.24), which can also be used to determine whether individuals are closely related.

Extension

In **Extension 5.3** you can compare the heterozygosity indices for natterjack toad populations. **X5.03S**

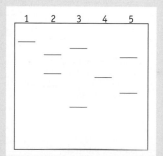

Figure 5.24 When DNA is cut with restriction enzymes some sequences produce fragments of different lengths, indicating that the individual is heterozygous for that section of the DNA. (See Topic 6 pages 73–8 for more detail on DNA fingerprinting.) On this DNA fingerprint five probes have been used for different DNA sequences. Sequences 1 and 4 give only a single band; the individual is homozygous for these sequences. The others have two bands, showing that the individual is heterozygous for these sequences.

5.4 What do we mean by ecological diversity?

Take a walk through a local wildlife area and you will soon see that the area is not uniform. A moorland, for example, is made up of undulating heather-covered slopes, streams, bogs, rocky outcrops and ponds, each habitat supporting a distinctive combination of organisms adapted to the particular conditions in that situation. There is a great deal of ecological diversity.

What is an ecosystem?

The part of the Earth and its atmosphere which is inhabited by living organisms is called the **biosphere**. Within the biosphere there are numerous different **ecosystems**. Ecosystems have distinctive features that affect the organisms living there. In an ecosystem there is the **abiotic** component – physical and chemical factors like the climate and soil type – and the **biotic** component – factors determined by organisms, such as **predation** and **competition**. It can be difficult to define the extent of an ecosystem but, in principle, ecosystems tend to be fairly self sustaining. So a lake remains as a lake (at least for a few hundred years), a wood remains as a wood, and so on.

The bottle of brine shrimps shown in Figure 5.25 is an ecosystem, although an unusually small and artificial one. It contains a community of organisms which interact with each other and with their physical environment in such a way as to make up a self-sustaining system (more or less!). The small crustaceans (the brine shrimps) in the bottle feed on the microscopic algae; the algae never run out but multiply as fast as the brine shrimps eat them; the carbon dioxide respired by the shrimps is taken up in algal **photosynthesis**; and so on.

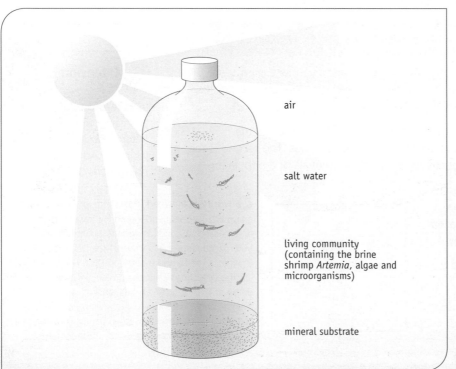

air

salt water

living community
(containing the brine
shrimp *Artemia*, algae and
microorganisms)

mineral substrate

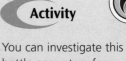

Activity

You can investigate this bottle ecosystem for yourself in **Activity 5.9**. **A5.09S**

◄ **Figure 5.25** A brine shrimp bottle ecosystem.

Q5.12 For the brine shrimp bottle:
a list some of the populations making up the biotic component of the ecosystem
b describe what makes up the abiotic component.

Habitats

Within an ecosystem there can be many different **habitats**. A habitat can be thought of literally as the place where an organism lives, with a distinct set of conditions. Within a pond ecosystem some organisms will live on the water surface, some anchored to the banks, others floating or swimming in the water. Each of these places constitutes the habitat of those organisms. Within a habitat there may be many tiny microhabitats, again each with distinct conditions. For example, the underside of a stone on the bottom of the pond will have a different set of conditions from the upper surface.

Communities

Within a habitat there will be several to many populations of organisms. Each **population** is a group of individuals of the same species found in an area. The various populations sharing a habitat or an ecosystem make up a **community**. So, for example, a woodland community consists of all the organisms that live in a wood.

Different niches avoid competition

If two organisms live in the same habitat and have exactly the same role within the habitat – the same food source, the same time of feeding, the same shelter site, and so on – they occupy the same **niche** and they will compete with each other. If one of the organisms is more successful, it may out-compete the other and exclude it from the habitat. Two organisms can occupy the same habitat using the same food source but have different niches (Figure 5.26).

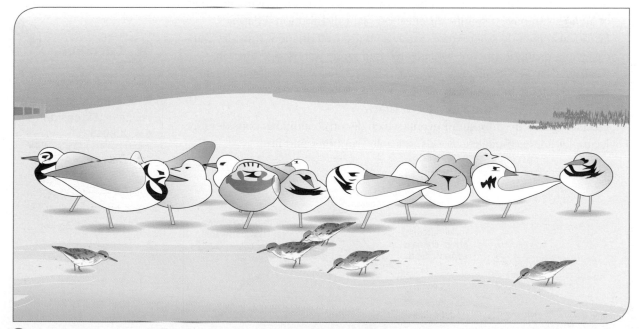

△ **Figure 5.26** More than one species can feed on the strand line using the same food source as long as they feed at different times. This ensures the different species occupy different niches.

What determines which species will occur in a particular habitat?

The reason why a particular species is found living in a particular place (unless deliberately introduced and maintained by humans) is because it happens to be there and can survive in the conditions of that site. Woods, moorland, fens, mud flats and meadows may look peaceful but they are all the scene of a 'dog-eat-dog' struggle for survival in which only the fittest survive. Species that come second don't gain a silver medal – they face extinction. The conditions in which species compete for survival are defined by a complex set of interrelated ecological factors. As we have seen, ecological factors can be roughly classified into abiotic and biotic ones.

Abiotic factors

Abiotic factors are non-living or physical factors; they include the following.

- **Solar energy input** is affected by latitude, season, cloud cover, and changes in the Earth's orbit. Light is vital to plants as it is the energy source for photosynthesis. It also has a role in initiating flowering, and in some species it is required for seed germination. In many animals, light affects behaviour. For example, changes in day length can be a cue for reproduction.
- **Climate** includes rainfall, wind exposure, and extremes of temperature.
- **Topography** includes altitude (which affects climate), slope, aspect (which direction the land faces) and drainage (see Figure 5.27).
- **Oxygen availability** is particularly important in aquatic systems. For example, fast-flowing streams are often better oxygenated than stagnant pools.
- **Edaphic** factors are connected with the soil, and include soil pH and mineral salt availability – itself affected by geology. The underlying geology of an area can have a significant effect on plant distribution. Edaphic factors also include soil texture. Sandy soils are well drained; they dry out easily in a drought, but are well aerated and rarely waterlogged in wet weather. Clay gets easily waterlogged, but it retains water well which can be an advantage in a drought.
- **Pollution** can be of the air, water or land.
- **Catastrophes** are infrequent events which disturb conditions considerably. Examples are earthquakes, floods, volcanic eruptions, and fires.

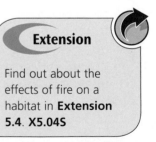

Extension

Find out about the effects of fire on a habitat in **Extension 5.4. X5.04S**

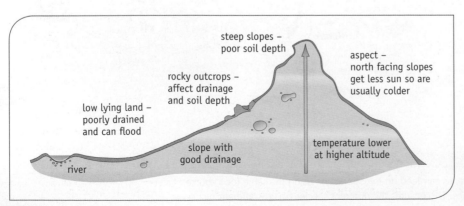

◀ **Figure 5.27** Topography affects conditions in a habitat.

Biotic factors

Biotic factors are 'living' factors. These include the following:

- **Competition** for resources such as food, light, water and space can be **interspecific** (between species) or **intraspecific** (within species).
- **Grazing**, **predation** and **parasitism** are relationships between two organisms where one benefits at the other's expense.
- **Mutualism** is a relationship in which both partners benefit.

Biotic factors are usually **density dependent**: the effects are related to the size of the population relative to the area available. The larger this population density, the greater the competition for food, space and so on.

There is normally a complex interaction of biotic and abiotic factors within a habitat. For example, if poor weather reduces the survival rate of one animal species, this has a knock-on effect on all its **predators**. If the pH of the soil changes, this may affect the survival of bacteria in the soil and the rate of decomposition and recycling of material may alter. Trees affect such abiotic factors as the water content of the soil and the humidity and light available for smaller plants. In turn, these abiotic factors influence other organisms in the ecosystem.

Anthropogenic factors

Anthropogenic factors are those arising from human activity. They can be either abiotic or biotic but it is important to recognise that the impact of humans on the world environment, our ecological footprint, is far greater than that of any other species.

Pollen records (as we reviewed in Topic 4) can reveal information about past vegetation. Such records show that when the Romans conquered Britain almost 2000 years ago it was largely forested. The landscapes of much of Britain would still be largely wooded were it not for deforestation, moor-burning and grazing. Grazing might be regarded as just a biotic factor since wild animals such as deer and rabbits also graze. But the introduction of sheep and rabbits and the removal of predators of grazing animals, such as wolves, are all the result of human actions. Grazing by domesticated animals such as sheep is also accompanied by high stocking densities, fencing, the introduction of cultivated types of grass, and the use of fertilisers. The result is that the environment is no longer 'natural'.

Q5.13 Look at the abiotic and biotic factors listed opposite and decide which are also anthropogenic.

Q5.14 Look at Figure 5.28 and list the major abiotic factors affecting this habitat.

Adapted for survival

Species survive in a habitat because they have adaptations that enable them to cope with both the biotic and abiotic conditions.

For example, the aye-aye in Figure 5.4 (page 3) has adaptations that aid its survival – a very long middle finger used for snagging grubs out of crevices in wood, long toes which aid climbing and large eyes useful in its nocturnal activities.

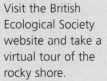

Weblink

Visit the British Ecological Society website and take a virtual tour of the rocky shore.

Activity

Activity 5.10 lets you study the relationship between the distribution of plant species and variations in abiotic factors. It includes a case study involving light intensity, soil pH and soil moisture, and considers some of the adaptations that enable particular species to survive in some places but not others. **A5.10S**

Sublittoral zone
- rarely uncovered except at extreme tide so conditions relatively constant
- wave action intense
- low light due to depth of water

Low shore
- relatively stable conditions
- exposed only at low tides so plenty of nutrients available
- less light penetrates water at high tide

Middle shore
- submerged for half the day so less desiccation than upper shore

Upper shore
- submerged only for short periods
- high desiccation
- wide variation in temperature and salinity

Splash zone
- rarely submerged
- extreme temperature variation and desiccation
- high salt content from spray
- dead organic matter accumulates on strand line

high water mark

low water mark

Laminaria species (brown alga)
- blade (or lamina) often found washed up on beaches because shed each autumn to reduce surface area
- holdfast attaches alga firmly to rocks
- adapted to low light intensities; additional photosynthetic pigments absorb maximum light

Chondrus crispus (red alga)
- grows around base of *Laminaria*
- shade tolerant
- adapted to low light intensities; additional photosynthetic pigments absorb maximum light

Blue rayed limpet
- herbivore
- feeds on *Laminaria*

Toothed wrack (brown alga)
- thin cell walls, intolerant of desiccation
- additional photosynthetic pigments absorb maximum light
- flat fronds in layers prevent the lower ones from drying out

Corallina (red alga)
- grows below brown algae or in rock pools
- shade tolerant,
- intolerant of desiccation

Dog-whelk
- carnivore
- on exposed shores develops a larger muscular foot to prevent being dislodged by waves
- shell size and thickness depend on wave action

Bladder wrack (brown alga)
- air bladders float fronds towards the light
- intolerant of desiccation

Flat periwinkle
- herbivore; feeds on wrack
- not tolerant of temperature variation
- several different colours of shell provide good camouflage against bladders of wrack
- eggs laid in a gelatinous mass to prevent drying out

Limpet
- herbivore
- adheres to rock with a muscular foot; clamps down onto rock to prevent drying out
- reduced metabolism when exposed

Channel wrack (brown alga)
- rolled fronds trap water and reduce water loss
- oily layer on surface of fronds slows desiccation
- thick cell walls shrink with drying
- survives low nutrient input
- grows very slowly

Spiral wrack (brown alga)
- spiralling fronds trap water
- thinner cell walls and no oily layer so loses water faster than channel wrack; often occurs lower on the shore
- shades out channel wrack

Rough periwinkle
- herbivore
- tolerant of high temperatures and desiccation by cementing itself to the rock and respiring without oxygen
- gills modified to absorb oxygen from the air

Lichens
- made up of a fungus and an alga or blue-green bacterium which have a mutualistic relationship; the algae or bacteria are protected from desiccation by the fungus and they provide the fungus with photosynthetic products

Black periwinkle
- grazes on lichens
- found in crevices
- tolerant of very variable salt and temperature conditions
- gill cavity modified to form a lung

Greater diversity of organisms as conditions become more stable and more nutrients are available.

▲ **Figure 5.28** A few examples illustrate how the distribution of organisms on the rocky shore is affected by biotic and abiotic factors, and show how species are adapted to conditions on the area of shore where they are found.

The familiar daisy not only survives grazing but relies on it. It is well adapted to survive close grazing and mowing. Can you spot the features in Figure 5.29 that help it cope?

Growing so close to the ground means that it cannot compete with tall plants for light, and it becomes less abundant in a lawn or meadow if the grass is allowed to grow tall. Daisies need grazing animals or mowers to keep down the opposition. Many of the species found in grazed or mown grassland have similar adaptations. This includes the grasses, which (like your hair) grow from the base and so can cope with being grazed.

Q5.15 What other adaptations do some plants have to escaping grazing?

Plants are not the only organisms well adapted to cope with grazing. The **herbivores** themselves are adapted. For instance, rabbits' teeth grow continually so that as they wear down new grinding surfaces emerge. Rabbits also eat their own pellets (faeces); grass is very hard to digest and, to gain maximum nutrition from their food, rabbits digest their food twice. They produce soft pellets the first time food passes through their gut; these are immediately eaten to digest the material further. Hard pellets are then produced as waste.

▲ **Figure 5.29** The daisy's growing point is at ground level, and its leaves form a basal rosette. Long roots enable it to find water when the soil gets dry (as often happens in grazed grassland on shallow soil).

Q5.16 Rabbits are not only well adapted for a diet of grass, they also have adaptations which help them avoid predators. What features of a wild rabbit help to assist its escape from predators?

Q5.17 Look at the organisms described in Figure 5.28. For one species in each trophic level, describe how it is well adapted to survive in the area of shore where it is found.

The more varied the environmental conditions within an ecosystem, the more numerous the different habitats, microhabitats and niches that organisms with suitable adaptations can occupy, creating greater biodiversity.

5.5 Biodiversity relies on energy transfer

Producers and productivity

The rate at which energy is incorporated into organic molecules in an ecosystem is called the **primary productivity** of the ecosystem. The primary productivity of terrestrial vegetation is usually positively correlated with plant diversity. Productive ecosystems, for example tropical rainforests, have a greater diversity of plants than ecosystems with low primary productivity.

Animal diversity is often linked to plant productivity, so that there is higher animal diversity in ecosystems with higher primary productivity. However, there is not always a positive correlation between plant productivity and animal diversity. For example, in lakes with high nutrient input (eutrophic lakes), there is increased algal productivity but a lower diversity of animals. This is due to the changing conditions accompanying **algal blooms**, principally depletion of oxygen as algae die and decompose.

Producers, also known as **autotrophs**, are organisms that can make their own organic compounds from inorganic compounds. Green plants, algae and some bacteria are producers. The algae in the bottle ecosystem (Figure 5.25) are producers. When light falls on producers, some energy is transferred to a chemical energy store by producing organic fuels such as glucose.

Not all primary producers are photosynthetic. Some are **chemosynthetic autotrophs**. These make organic molecules using energy released from chemical reactions (Figure 5.30).

Photosynthesis

△ **Figure 5.30** Chemosynthetic bacteria near volcanic vents in the deep ocean supply energy to food chains without using light.

An overview

In photosynthesis carbon dioxide is reduced as hydrogen and electrons from water are added to it, creating a carbohydrate. The overall equation is:

$$6CO_2 + 6H_2O \xrightarrow[\text{in the presence of chlorophyll}]{\text{energy from light}} C_6H_{12}O_6 + 6O_2$$

The reactions in photosynthesis require an input of energy from light. The energy needed to break the bonds within carbon dioxide and water is greater than the energy released when the products – glucose and molecular oxygen – are formed. Therefore, the products of the reaction (glucose and oxygen) are at a higher energy level than the reactants (carbon dioxide and water), and act as a store of energy. You can think of this as being similar to stretching (or compressing) a spring to store energy. In photosynthesis, oxygen is a waste product and is released into the atmosphere. Glucose is a fuel, which can later be oxidised during respiration to release energy.

If you have ever seen hydrogen burned in air you will be aware that the reaction between hydrogen and oxygen releases large amounts of energy. In photosynthesis the hydrogen is separated from water and stored within a carbohydrate. When energy is required in the cell the stored hydrogen reacts with oxygen during respiration, releasing a large amount of energy.

Releasing hydrogen from water

The splitting of water into hydrogen and oxygen requires energy. Photosynthesis uses energy from sunlight to split water. This process is known as the **photolysis** of water because of the involvement of energy from light in the reaction (*photo* means 'light'; *lysis* means 'splitting').

Storing hydrogen in carbohydrates

The hydrogen is reacted with carbon dioxide to 'store' the hydrogen as an organic fuel. Carbon dioxide is reduced to form glucose, which can be stored or converted to other organic molecules. This fuel then has the potential to release large amounts of energy when the hydrogen stored in the organic fuel reacts with oxygen during respiration. In aerobic respiration glucose is pulled apart, the hydrogen combining with oxygen to make water, and carbon dioxide is released.

Photosynthesis converts carbon dioxide to carbohydrates using energy from light and hydrogen from water.

Photosynthesis is vital for the life of our planet. The organic fuels made during photosynthesis are then passed on through food webs to other organisms in the ecosystem. These molecules provide fuels for energy and raw materials for synthesis of a wide range of organic molecules. Provided that there is an external source of energy, everything else (e.g. minerals, water) within the ecosystem can be recycled.

How photosynthesis works

Photosynthesis is not a single reaction, but a series of reactions. These reactions are in two main stages (Figure 5.31).

- **Light-dependent reactions** use energy from light and hydrogen from photolysis of water to produce **reduced NADP**, **ATP** and the waste product oxygen. The oxygen is either used directly in respiration or released into the atmosphere.
- **Light-independent reactions** use the reduced NADP and ATP from the light-dependent reactions to reduce carbon dioxide to carbohydrates.

What is meant by reduction? Covalent bonding involves the electrons in atoms. When bonds break or new bonds form there are movements of electrons from one reactant to another. Electrons are also transferred when atoms turn into ions, or ions into atoms. The loss of electrons from a substance is known as oxidation and the gain of electrons is known as reduction.

Remember this using OIL RIG, **o**xidation **i**s **l**oss, **r**eduction **i**s **g**ain. When a substance is in an oxidised form it has lost electrons. In a reduced form a substance has gained electrons. The **coenzyme NADP** is reduced when hydrogen and electrons are added during photosynthesis.

Support

To find out more about energy transfer in chemical reactions look at the Biochemistry support on the website.

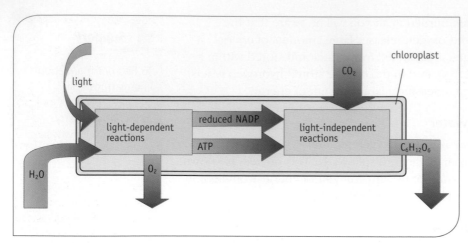

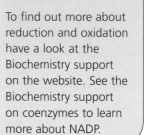

◀ **Figure 5.31** The two stages of photosynthesis.

Although the overall equation for photosynthesis suggests that carbon dioxide and water react with one another, the carbon dioxide and water never come into direct contact with each other. The hydrogen, electrons and energy needed for the reduction of CO_2 are transferred indirectly using reduced NADP and ATP (energy transfer molecule, see page 35).

Q5.18 Look at the summary of photosynthesis shown in Figure 5.31 and then answer the following questions.
a Which molecule provides the source of hydrogen for photosynthesis?
b What is the source of energy for photosynthesis?
c Which molecule provides the hydrogen for the light-independent reactions?
d Which molecule provides the energy for the reduction of carbon dioxide to glucose in the light-independent reactions?

Where does photosynthesis take place?

The site of photosynthesis is the **chloroplast**. A **palisade mesophyll** cell in a leaf can contain as many as 50 chloroplasts. Each chloroplast is made up of membranes, arranged in a very precise, organised way as shown in Figure 5.32.

The light-dependent reactions

When light is absorbed by the photosynthetic pigments in the **thylakoid membranes** of the chloroplast the following events occur:

1 Energy from the light raises two electrons in each chlorophyll molecule to a higher energy level. The chlorophyll molecules are now in an 'excited' state.

2 The electrons leave the excited chlorophyll molecules and pass along a series of **electron carrier** molecules, all of which are embedded in the thylakoid membranes. These molecules constitute the **electron transport chain**.

3 The electrons pass from one carrier to the next in a series of oxidation and reduction reactions losing energy in the process. The energy is used in the synthesis of ATP. Details of electron transport chains are covered in Topic 7.

4 The electrons lost from the chlorophyll must be replaced if photosynthesis is to continue and maintain the flow of electrons along the electron transport chain.

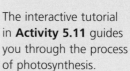

Support

To find out more about reduction and oxidation have a look at the Biochemistry support on the website. See the Biochemistry support on coenzymes to learn more about NADP.

Activity

The interactive tutorial in **Activity 5.11** guides you through the process of photosynthesis. **A5.11S**

A

Thylakoid membranes – a system of interconnected flattened fluid-filled sacs. Proteins, including photosynthetic pigments and electron carriers, are embedded in the membranes and are involved in the light-dependent reactions.

Stroma – the fluid surrounding the thylakoid membranes. Contains all the enzymes needed to carry out the light-independent reactions of photosynthesis.

Thylakoid space – fluid within the thylakoid membrane sacs contains enzymes for photolysis.

Granum – a stack of thylakoids joined to one another. Grana (plural) resemble stacks of coins.

A smooth outer membrane – which is freely permeable to molecules such as CO_2 and H_2O

A smooth inner membrane – which contains many transporter molecules. These are membrane proteins which regulate the passage of substances in and out of the chloroplast. These substances include small molecules like sugars and proteins synthesised in the cytoplasm of the cell, but used within the chloroplast.

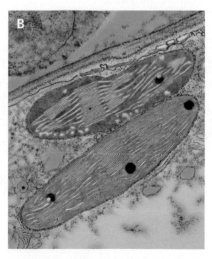

B

Figure 5.32 A The structure and function of chloroplasts. **B** Electron micrograph of two chloroplasts. Magnification ×16 000.

Weblink

Visit the interactive tutorial on cell structure and function in **Activity 3.1** to check out the structure of chloroplasts.

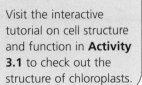

Extension

In **Extension 5.5** you can look in detail at the structure of the leaf. **X5.05S**

5   Within the thylakoid space, an enzyme catalyses the splitting of water (photolysis) to give oxygen gas, hydrogen ions and electrons. These electrons replace those that were emitted from the chlorophyll molecule, so it is no longer positively charged. The hydrogen ion concentration within the thylakoid is raised as a result of photolysis.

6   The electrons that have passed along the electron transport chain combine with the coenzyme NADP and hydrogen ions from the water to form reduced NADP.

The ATP and reduced NADP created in the light-dependent reactions are used in the light-independent reactions.

The light-dependent reactions are summarised in Figure 5.33.

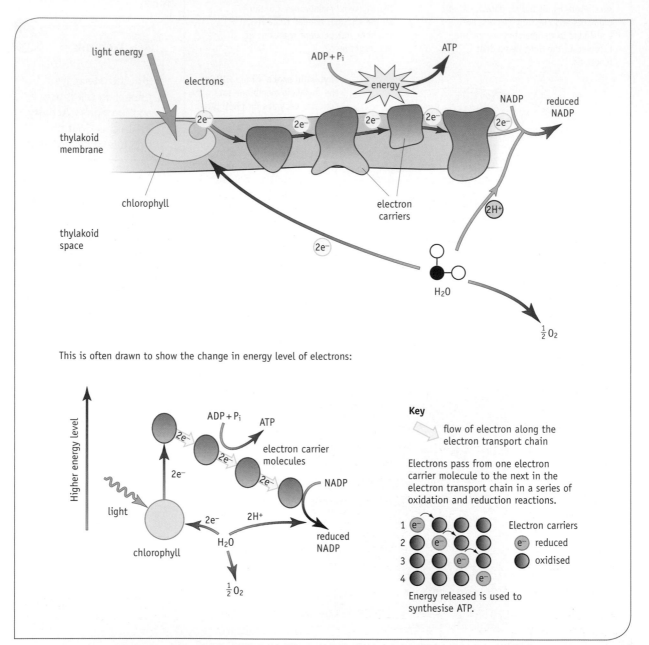

Figure 5.33 In the light-dependent reactions of photosynthesis, the energy from the Sun excites chlorophyll molecules. High energy electrons leave the chlorophyll and pass along a series of carrier proteins in the electron transport chain. The ionised chlorophyll causes the photolysis of water; hydrogen is pulled off water molecules raising the local hydrogen ion concentration within the thylakoid. ATP is also formed as a phosphate group is added to ADP.

Q5.19 Write equations summarising:
a the splitting of water
b the reduction of NADP.

Extension

You can learn more about the light-dependent reactions of photosynthesis in **Extension 5.6**. **X5.06S**

Key biological principle: The role of ATP

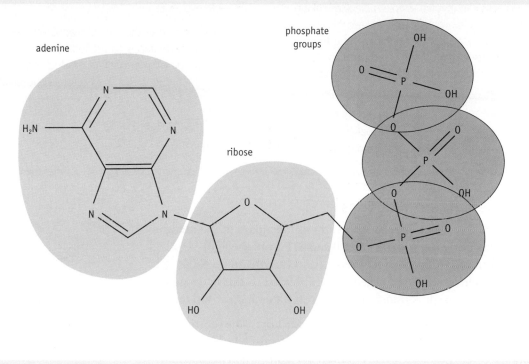

🔺 **Figure 5.34** The structure of ATP.

ATP (**adenosine triphosphate**) is the most important energy transfer molecule within cells. It moves energy around the cell from energy yielding reactions to energy-requiring reactions.

ATP consists of adenine (an organic base), ribose (a 5-carbon sugar) and three phosphate groups (Figure 5.34). The third phosphate group on ATP is only loosely bonded to the second phosphate, so is easily removed. When this phosphate group is removed from ATP, **adenosine diphosphate** forms: **ADP**. Once removed, this phosphate group becomes hydrated, forming bonds with surrounding water molecules. A lot of energy is released as bonds form between water and the phosphate group. This energy can be used to drive energy-requiring reactions in the cell. **ATPase** catalyses the breakdown of ATP to ADP.

ATP in water → ADP + hydrated P_i + energy

ATP is created from ADP by the addition of inorganic phosphate (P_i). In order to make ATP, phosphate must be torn away from water molecules, and this reaction requires energy. ATP in water is higher in energy than ADP and phosphate ions in water, so ATP in water is a way of storing chemical potential energy. Formation of ATP separates the phosphate and water. The phosphate and water can be brought together in an energy-yielding reaction each time energy is needed for reactions within the cell.

Support

To find out more about ATP visit the Biochemistry support on the website.

Reactions in cells can involve simple molecules linking to form more complex molecules. We saw this in the creation of polysaccharides in Topic 1 and the formation of proteins in Topic 2. Reactions can also involve the breakdown of complex molecules into simpler molecules.

Reactions rarely occur in a single step such as A → D. Instead there is generally a series of smaller reactions forming a metabolic pathway, for example:

A → B → C → D

This allows the rate of the overall reaction to be controlled, as each step is controlled by a specific enzyme. A range of intermediate products can be produced which might each be useful as end-products or take part in other reactions, for example:

A → B → C → D
↓
E → F

The light-independent reactions

The light-independent reactions of photosynthesis take place in the **stroma** of the chloroplasts using the reduced NADP and ATP from the light-dependent reactions. Carbon dioxide is reduced to carbohydrate. NADP acts as a hydrogen carrier, keeping hydrogen loosely bonded so this hydrogen can't react with oxygen as it is transferred between water and carbon dioxide.

As with all metabolic pathways, there is a series of reactions. The reactions form a cyclical pathway called the **Calvin cycle** after the scientist who first worked it out. A simplified version of the cycle is shown in Figure 5.35. This diagram shows how carbon dioxide combines with a 5-carbon compound. An unstable 6-carbon molecule forms that almost immediately breaks down into two 3-carbon compounds. Some of the 3-carbon compounds created are reduced to form carbohydrate, using the electrons and hydrogen from the reduced NADP created in the light-dependent reactions. Some of the 3-carbon compounds are used to regenerate the original 5-carbon compound that accepted the carbon dioxide. This recycling is a very economical use of resources by the cell.

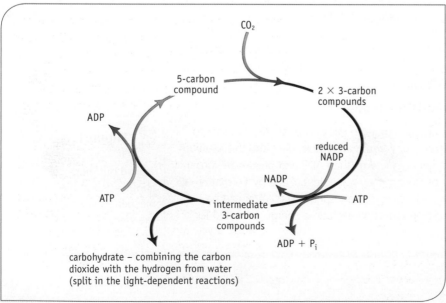

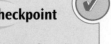

Figure 5.35 The key reactions that occur in the Calvin cycle.

Activity

In **Activity 5.12** you complete practical work to investigate photosynthesis reactions.

Checkpoint

5.3 Produce a flowchart or bullet-point summary that describes the steps in the light-dependent and light-independent reactions of photosynthesis.

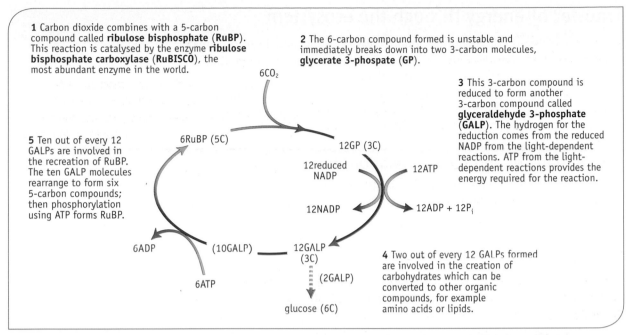

1 Carbon dioxide combines with a 5-carbon compound called **ribulose bisphosphate** (**RuBP**). This reaction is catalysed by the enzyme **ribulose bisphosphate carboxylase** (**RuBISCO**), the most abundant enzyme in the world.

2 The 6-carbon compound formed is unstable and immediately breaks down into two 3-carbon molecules, **glycerate 3-phospate** (**GP**).

3 This 3-carbon compound is reduced to form another 3-carbon compound called **glyceraldehyde 3-phosphate** (**GALP**). The hydrogen for the reduction comes from the reduced NADP from the light-dependent reactions. ATP from the light-dependent reactions provides the energy required for the reaction.

5 Ten out of every 12 GALPs are involved in the recreation of RuBP. The ten GALP molecules rearrange to form six 5-carbon compounds; then phosphorylation using ATP forms RuBP.

4 Two out of every 12 GALPs formed are involved in the creation of carbohydrates which can be converted to other organic compounds, for example amino acids or lipids.

$6CO_2$

6RuBP (5C)

12GP (3C)

12reduced NADP

12ATP

12NADP

$12ADP + 12P_i$

6ADP

(10GALP)

12GALP (3C)

(2GALP)

6ATP

glucose (6C)

Figure 5.36 shows the reactions of the Calvin cycle in more detail. Read the annotations to discover exactly what is happening at each stage of the cycle. Even this is a simplification. In reality there are a large number of intermediate reactions.

▲ **Figure 5.36** The light-independent reactions of the Calvin cycle (also known as the C3 pathway).

Q5.20 Look at Figure 5.36 which shows the light-independent reactions.
a Name a substance that has been phosphorylated (had phosphate added to it).
b Name a substance formed by reduction.
c Explain why the diagram shows six RuBP molecules combining with six carbon dioxide molecules rather than just one of each.

Key biological principle: Why does photosynthesis take place inside chloroplasts?

Thylakoids and the light-dependent reaction
Photosynthesis consists of a series of enzyme-controlled reactions in which some of the energy falling on the plant surface is initially stored as chemical potential energy within ATP in water. The formation of ATP occurs in the light-dependent reactions as a result of a series of oxidation and reduction reactions. These reactions involve the transfer of electrons between electron carrier molecules. The electron carriers are located within the thylakoid membrane in the chloroplast. Their positioning within the membrane creates an electron transport chain, allowing electrons to pass efficiently from each electron carrier to its neighbour.

Stroma and the light-independent reaction
The ATP molecules that are formed act as energy carriers within the cell, allowing small amounts of energy to be transferred and used where needed. In photosynthesis, the energy stored within ATP in water is used in reactions within the light-independent stage, in the fixing of carbon dioxide within organic molecules. These reactions are dependent on the collision of substrate(s) and the appropriate enzymes to catalyse the reactions.

Maintaining a high concentration of each enzyme throughout the cell would be very 'costly' in terms of synthesis of enzyme. On the other hand, low concentrations would reduce the rate of reaction and efficiency of photosynthesis. The compartmentalisation of these reactions within the chloroplast stroma means that the substrates and enzymes can be at concentrations which allow the reactions to be successfully catalysed as quickly as possible.

Transfer of energy through the ecosystem

The first photosynthetic organisms found a way of using sunlight to pull apart water molecules. Nowadays the Earth has an oxygen-rich atmosphere due to the oxygen released as a waste product from this reaction. In turn, the oxygen-rich atmosphere makes possible the reactions in aerobic respiration, which release large amounts of energy as hydrogen and oxygen are brought back together to make water.

Energy transfer and feeding relationships

Some of the energy fixed within organic molecules by autotrophs (also known as producers) is transferred to other organisms in the ecosystem. Organisms which obtain energy as 'ready-made' organic matter by ingesting material from other organisms are known as **heterotrophs**. Heterotrophs include all animals, all fungi, most bacteria and some protoctists.

Heterotrophs cannot make their own food; instead they must consume it. All heterotrophs are **consumers** and depend on producers (autotrophs) for their food.

- **Primary consumers**, also called **herbivores**, are heterotrophs which eat plant material.
- **Secondary consumers**, also called **carnivores**, feed on primary consumers.
- **Tertiary consumers** (also carnivores) eat other consumers. The carnivores at the top of the food chain are sometimes called top carnivores.

Animals that kill and eat other animals are known as predators and carnivores ('flesh eaters'). Animals that eat plants and other animals are known as **omnivores**.

Energy is transferred from producers to primary consumers, then to secondary consumers and then, sometimes, to tertiary consumers. Such feeding relationships can be shown in a **food chain** (Figure 5.37) or **food web**. The position a species occupies in a food chain is called its **trophic level**. Energy is transferred from one trophic level to the next trophic level.

Q5.21 In the brine shrimp food chain in Figure 5.37, which organism occupies the position of:

a producer
b primary consumer
c secondary consumer
d tertiary consumer?

Detritivores are primary consumers that feed on dead organic material called detritus. Woodlice, earthworms and freshwater shrimps are examples of detritivores.

Decomposers are species of bacteria and fungi that feed on the dead remains of organisms and on animal faeces. Like animals, they are heterotrophs. They secrete enzymes and digest their food externally, before absorption takes place.

Activity

You can construct a food web and investigate trophic levels in **Activity 5.13**. **A5.13S**

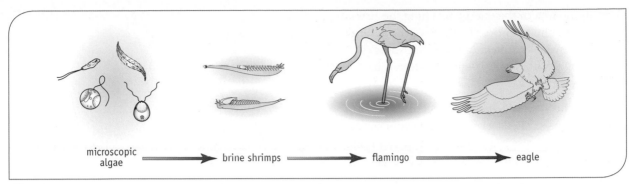

Figure 5.37 The brine shrimp food chain. Most ecosystems have complex food webs in which each organism eats or is eaten by several other organisms.

microscopic algae ⟶ brine shrimps ⟶ flamingo ⟶ eagle

Decomposers and detritivores play an important role in the recycling of organic matter from dead remains and waste. In the bottle ecosystem (Figure 5.25) bacteria living in the bottle cause the decomposition of dead brine shrimps and brine shrimp droppings. The organic compounds are broken down into inorganic substances that are taken up by the algae. The algae can grow and reproduce, replacing those eaten by brine shrimps.

How efficient is the transfer of energy through the ecosystem?

The productivity (and to some extent species biodiversity) of an ecosystem will depend on how much energy is captured by the producers and how much is transferred to the higher trophic levels.

Plants are not actually as good as you might think at absorbing light, and only a small fraction of the light energy that reaches a plant is used in photosynthesis. It is estimated that 1×10^6 kJ m^{-2} year^{-1} of energy is intercepted by plants in the UK, but that less than 5% of this available energy is captured in photosynthetic products.

For a start, if you look at Figure 5.38 you can see that most energy reaching the plant is not even absorbed. This is largely because chlorophyll can only absorb certain wavelengths, as shown by the absorption spectrum in Figure 5.39.

Checkpoint

5.4 Define the terms:
- habitat
- population
- community
- niche
- ecosystem
- autotroph
- heterotroph
- producer
- primary consumer
- secondary consumer
- predator
- trophic level
- decomposer.

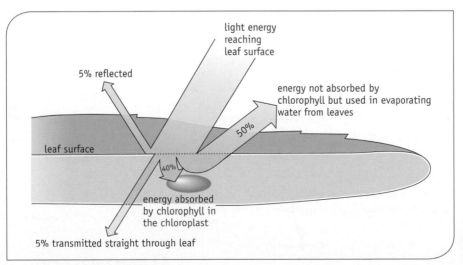

light energy reaching leaf surface

5% reflected

energy not absorbed by chlorophyll but used in evaporating water from leaves

50%

leaf surface

40%

energy absorbed by chlorophyll in the chloroplast

5% transmitted straight through leaf

Figure 5.38 Not all the light energy falling on the leaf gets absorbed by the chlorophyll.

Only about 40% of the energy reaching the leaf is absorbed by the chlorophyll. Much of this energy is used to make organic molecules but some is lost during photosynthesis and transferred to the environment.

Limiting factors will also influence the rate of photosynthesis. The law of limiting factors states that when a process is affected by more than one factor, its rate is limited by the factor furthest from its optimum value. So, photosynthesis can be limited by temperature in cool conditions and by light in overcast conditions. In the UK during the summer the main limiting factor is carbon dioxide concentration. This means that not all the light energy falling on the plant surface can be used to make organic molecules even if it is absorbed by the photosynthetic pigments.

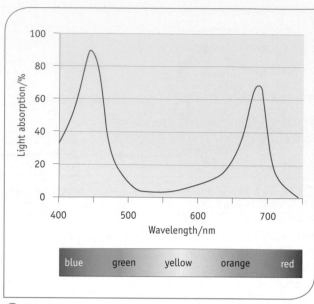

▲ **Figure 5.39** Absorption spectrum for chlorophyll.

Q5.22 Look at Figure 5.39. Which parts of the visible spectrum are absorbed by chlorophyll?

The rate at which energy is incorporated into organic molecules by an ecosystem is the **gross primary productivity** (**GPP**). GPP is usually expressed as units of energy per unit area per year (e.g. kJ m^{-2} y^{-1} or MJ ha^{-1} y^{-1}).

The percentage efficiency of photosynthesis can be calculated as GPP divided by the amount of light energy striking the plant times 100. Very little of the light energy that reaches a plant is fixed in carbohydrates. For example, in a grassland community where 2×10^9 kJ m^{-2} y^{-1} of sunlight energy reaches the plant and GPP is 25×10^6 kJ m^{-2} y^{-1}, the efficiency of photosynthesis is 1.25%.

Some of the carbohydrates produced are quickly broken down in respiration. This provides the energy for the plant's life processes, such as cell division and active transport. The rest of the carbohydrates are incorporated into the proteins, chromosomes, membranes and other components of new cells, becoming new plant **biomass**. The rate at which energy is transferred into the organic molecules that make up the new plant biomass is called **net primary productivity** (NPP), Figure 5.40.

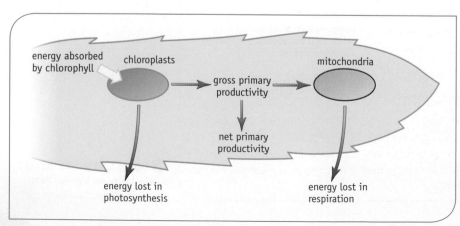

◀ **Figure 5.40** Transfer of energy absorbed by chlorophyll.

Net primary productivity, gross primary productivity and respiration are related to each other by the equation:

$$NPP = GPP - R$$

where R is plant respiration.

The net primary productivity is available to the rest of the ecosystem.

Q5.23 **a** Using the figures in Table 5.3 calculate the missing values **i–iii** for each of the ecosystems. Comment on the data.
b 2×10^6 kJ of sunlight energy falls on each square metre of the maize field described in Table 5.3. Calculate the efficiency of photosynthesis.
c Explain why so little of the sunlight energy is used in photosynthesis.
d A desert was found to have a NPP of 836 kJ m^{-2} y^{-1}. Suggest reasons why this value is so low.

▼ **Table 5.3**

Ecosystem	GPP/ kJ m^{-2} y^{-1}	Respiration/ kJ m^{-2} y^{-1}	NPP/ kJ m^{-2} y^{-1}
tropical rainforest	23 140	17 820	i
young pine forest	5100	1960	ii
maize field	iii	8000	26 000

▲ **Figure 5.41** Will this pine forest ecosystem or a tropical rainforest ecosystem have the higher NPP? Complete Question 5.23 to find out if you were right.

Disappearing energy

Producers to primary consumers

Transfer of energy from producers to primary consumers is also not very efficient. Only about 2–10% of the energy in the producers goes to make new herbivore biomass. What happens to the rest?

- *Not all the available food gets eaten.* This may be due to limitations of the animals' feeding methods. When a bull eats grass, for example, it can only eat long untrampled grass because it needs to wrap its tongue around the grass and rip it up. Some parts of the plants, for example the roots, twigs and parts protected by spines or thorns, will not get eaten by herbivores.
- *Some energy is lost in faeces and urine.* The main component of plant material, apart from water, is the cellulose of the cell walls. Cellulose is tough stuff to deal with, and mammals have no enzymes of their own to help break it down. Even in ruminants, animals whose guts contain microbes producing cellulase enzymes, much of the cellulose still passes through the gut intact and comes out in the faeces.
- *Much of the energy absorbed by the consumers is used in respiration* for movement and chemical reactions in the body, and is lost to the environment as heat.

Primary consumers to secondary consumers

The transfer of energy from primary to secondary consumers – from herbivores to carnivores – is often more efficient. Often, over 10% of the energy in herbivores ends up in carnivores because a high proportion of the herbivores may be eaten by carnivores and the protein-rich diet is easily digested so there is less lost in faeces.

The fate of energy within a trophic level

The energy entering a trophic level must equal the amount used or lost by that trophic level. We can summarise this in a simple equation:

$$\begin{array}{l} \text{energy entering the trophic} \\ \text{level (i.e. consumed)} \end{array} = \begin{array}{l} \text{energy lost in respiration + energy lost in faeces} \\ + \text{ energy lost in urine + energy in new biomass} \end{array}$$

Q5.24 Look at the cow in Figure 5.42 and work out:
a How much energy is transferred to new biomass?
b What percentage is this of the energy consumed?

Energy flow and energy 'loss' provide the explanation of why food chains and food webs rarely have more than four or five trophic levels.

What always gets less and less as you go up a food chain is *the transfer of energy to the next trophic level*. If a diagram is drawn with a bar representing the energy transferred from one trophic level to the next, in kJ m^{-2} y^{-1}, a pyramid is formed. The bars representing the higher trophic levels are always smaller (Figure 5.43). Such pyramids allow comparisons of the efficiency of energy transfer between trophic levels.

Q5.25 Look at Figure 5.43 and work out the efficiency of transfer between:
a trophic level 1 and trophic level 2
b trophic level 2 and trophic level 3.

▼ **Figure 5.43** A pyramid of energy for a grazed pasture (drawn to scale). Each bar represents the energy transferred from one trophic level to the next, in kJ m^{-2} y^{-1}. The bottom bar represents the energy from sunlight captured by producers in trophic level 1; the second bar is the energy transferred from level 1 (producers) to the primary consumers in trophic level 2; and so on.

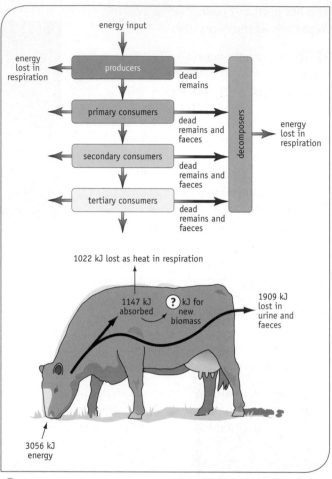

▲ **Figure 5.42** The flow of energy through the ecosystem.

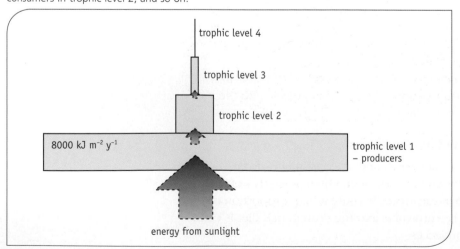

Activity

Activity 5.14 lets you work out net primary productivity for an ecosystem and explore the transfer of energy through the ecosystem. **A5.14S**

5.6 What caused the huge variety of life?

Until the eighteenth century it was considered that all species had been created pretty much exactly as they then existed. But biologists gradually became less convinced of this, partly because they kept on finding fossils of extinct species.

Evolution by natural selection

The Origin of Species

In 1859 Charles Darwin published his book *On the Origin of Species by Means of Natural Selection or The Preservation of Favoured Races in the Struggle for Life*. This set out his theory that organisms might have evolved, gradually changing from one form to another.

One of the things that greatly influenced Darwin in the development of his ideas in *The Origin of Species* was the great diversity of life he saw on the voyage of the *Beagle*. He was particularly struck by what he saw on the volcanic islands of the Galapagos Islands where he observed closely related species. These were clearly similar to, but not the same as, species on the South American mainland. One example that influenced his thinking was the giant land tortoises (Figure 5.44). He was told at a dinner on one of the islands that these tortoises had a different shaped shell on each island.

△ **Figure 5.44** A giant tortoise on the Galapagos Islands.

Darwin's interest in breeding fancy pigeons made him aware of the enormous variation within a species which could be generated by selective breeding. What he struggled with was how that variation could be inherited, as he had no knowledge of genetics.

Another great influence on Darwin's thinking was Malthus and his ideas about the factors influencing the size of populations. In 1798, Malthus wrote an essay about the potential of the human population to grow exponentially. He argued that this would lead to a population size that would exceed available food supply, resulting in famine, pestilence and war. Darwin realised that the sizes of many animal populations were staying the same.

He had also read the work of the geologist Charles Lyell who suggested that the world was many millions of years old, rather than mere thousands of years as was widely believed at the time.

Darwin put these ideas and observations together and came up with the theory of **evolution** by **natural selection**. For a long time Darwin did not publish his findings. But then another naturalist, Alfred Russell Wallace, who had also read Malthus and had observed the variety of life in South America and Malaysia, sent Darwin an essay outlining a theory that was identical to Darwin's own. They published a short paper together outlining the theory. This stimulated Darwin to put together the mass of material he had gathered in support of the theory and he finally published *The Origin of Species* in 1859, 28 years after joining the survey ship HMS *Beagle*.

Activity

Activity 5.15 lets you investigate evolution. **A5.15S**

A summary of Darwin's theory is as follows:
- *Darwin observed:* Organisms produce more offspring than can survive and reproduce. Numbers in natural populations stay much the same over time.
- *He concluded:* There is a **struggle for existence** – competition for survival between members of the same species. As a population increases in size, environmental factors halt the increase (see Figure 5.45). Many individuals die due to predation, competition for food and other resources, or due to the rapid spread of disease resulting from overcrowding.

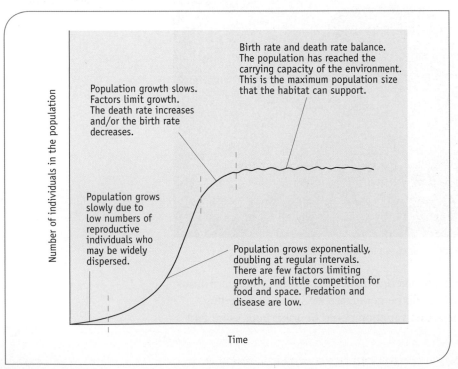

Figure 5.45 Population growth.

- *Darwin also observed:* There is a huge amount of variation within species.
- *He concluded:* Those individuals that are best adapted to conditions in their environment are more likely to survive and breed. They have a selective advantage: natural selection is acting, and there is **survival of the fittest**. Individuals with these adaptive features will be more common in the next generations. Those organisms that are not well adapted are more likely to die before maturity and so do not produce many offspring. Over a period of time, the character of the species will change to the more adapted form.

Lamarck's theory

Darwin and Wallace's theory of natural selection contrasts with the earlier theory put forward by the French naturalist Jean Baptiste de Lamarck in 1809. He noted that structures or organs that are constantly used become well developed. For example, blacksmiths develop large arm muscles, whereas muscles that are not used shrink. Lamarck suggested that over several generations structures could change in response to the environment, as the environment of an organism will determine whether an organism needs to use the structure. This he termed the 'first law'.

He proposed a 'second law', namely that the changes acquired by an organism were inherited; thus a blacksmith's children would inherit larger arm muscles. The classic example used to illustrate his ideas is the neck of the giraffe (Figure 5.46). This would have elongated due to the need to reach food in trees. The slightly longer neck that resulted would be passed on to offspring and the process would be repeated, with the neck gradually lengthening over the generations.

▲ **Figure 5.46** How did the giraffe's long neck come about?

Lamarck based his theory on the available knowledge at the time. Given that biologists knew nothing of genes and the mechanisms of inheritance, Lamarck's theory is plausible but has been superseded by that of Darwin and Wallace's theory of natural selection, which is supported by a great deal of evidence.

Q5.26 **a** Explain how the long neck of the giraffe would have come about according to Darwin and Wallace's theory of natural selection.
b Explain why acquired traits are extremely unlikely to be inherited.

The theory today

Today, the theory of evolution by natural selection remains much as it was first proposed by Darwin and Wallace, although with our knowledge of genetics and inheritance we now think of natural selection as acting on the alleles or groups of alleles which are responsible for the inherited variation. Thus evolution is a genetic change in a population of organisms over time (generations). The fundamentals of the theory are summarised in the five points below.

1 A population has some naturally occurring genetic variation with new alleles created through mutations.

2 A change in the environment causes a change in the selection pressures acting on the population.

3 An allele that was previously of no advantage now becomes favourable.

4 Organisms with the allele are more likely to survive, reproduce and so produce offspring.

5 Their offspring are more likely to have the allele, so it becomes more common in the population.

Natural selection is happening right now!

Natural selection is not something that happened only in the past, or something that takes millions of years. It is happening all the time. Natural selection tends to keep things the same, as Figure 5.47 shows. But if conditions change then evolution occurs. For example, natural selection has resulted in evolution within head lice (Figure 5.48).

Extension

You can find out how to calculate allele frequencies in **Extension 5.7**. X5.07S

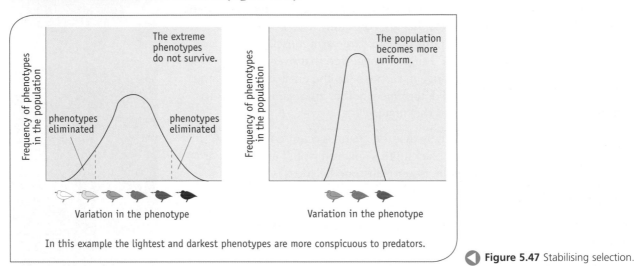

In this example the lightest and darkest phenotypes are more conspicuous to predators.

Figure 5.47 Stabilising selection.

About 7% of UK primary school children have an infestation of head lice in their hair. These parasites suck blood and have a rapid rate of reproduction, doubling in number every seven days. There are many head lice shampoos which contain insecticides. Permethrin and malathion are frequently the active ingredients. However, in recent years there is both anecdotal and scientific evidence to show that head lice have become resistant to these shampoos which are no longer as effective. Results of a study carried out at the University of Bristol showed that there was an 87% failure rate for permethrin and a 64% failure rate for malathion in the treatment of infested school children.

This is a modern example of natural selection in action. Strictly, it's an example of artificial selection – that is, selection as a result of human activity, though the selection is unintentional. Initially, many head lice must have been killed by the insecticides, but now it seems that permethrin will kill only 13% of them. Most head lice survive and quickly breed the next resistant generation. The processes that have brought this about are presumably something like the following.

▶ **Figure 5.48** Coloured scanning electron micrograph (SEM) of a human head louse *Pediculus humanus*. An egg (nit) has been laid on the hair (lower right). The lid of the nits (top) are perforated with air holes that allow the developing nymphs to breathe. Adult female lice lay between 80 and 100 eggs in a lifetime. Magnification ×38.

- Variation in the population – there were some individuals in the louse population with an allele giving them the ability to metabolise the insecticide, rendering it harmless. This allele arose from a mutation in the gamete-producing cells, which meant it could be inherited.
- Selection pressure – on a head of hair treated with insecticide, the head lice with the 'normal' allele will be killed, leaving the surviving resistant head lice with fewer competitors.
- Fast life cycle – a life cycle of approximately three weeks from egg to mature adult leads to a rapid increase in the number of resistant lice.

In this example, the insecticides in the shampoos provide the selection pressure. Had there not been pre-existing variation in the population, all the lice on treated heads would have died. Without that original chance mutation the head louse could have become extinct – at least in areas where the chemical treatments against it were used.

Evolution has occurred: the insecticide-resistance allele in head lice went from being quite rare in pre-insecticide populations to being extremely common.

In the case of the resistant head lice, evolution by natural selection has occurred producing a population which has different allele frequencies from the original population, but it does not produce separate species. The insecticide-resistant lice can still breed with the non-resistant ones, so do not count as a separate species. Some additional mechanism(s) must be involved in the formation of new species.

Q5.27 The pictures in Figure 5.49 on the next page show the events in a hypothetical instance of evolution. Match statements 1 to 5 on pages 45–6 with the correct pictures.

Speciation

How are new species formed?

The formation of a new species is called **speciation**. Darwin thought that, over time, populations that were exposed to selection would become so different from their original form that they would eventually become different species. It is now generally accepted that, for a new species to arise, a group of individuals has to be reproductively isolated from the rest of the population.

The most common method of speciation is thought to be isolation of part of a population by some geographical feature which prevents a group of individuals from breeding with the rest of the population, such as a high mountain range, a river, or a stretch of ocean (Figure 5.50). Over time, the two groups will become less like each other as they respond to different selection pressures within their local habitats, and as random mutations accumulate. When the members of the two groups meet again, they may not be able to interbreed if these differences are great enough.

Once the two populations are unable to breed and produce fertile offspring they are considered to belong to two different species, and there is **reproductive isolation** between them. The longer the two groups are geographically isolated, the more likely it is that speciation will occur.

Checkpoint

5.5 a Write a summary explaining how natural selection can lead to evolution.
b Produce a time line detailing the historical development of the theory of evolution.

▲ **Figure 5.49** Evolution in action.

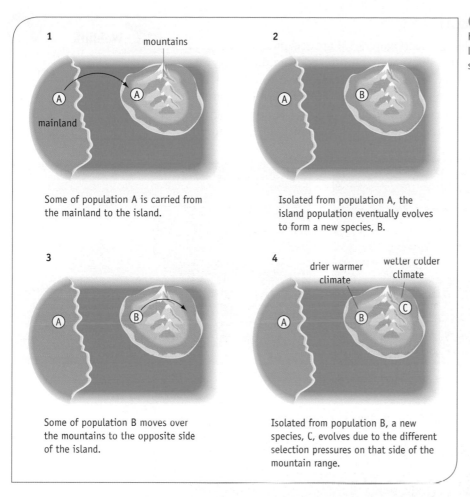

1

mountains

A → A

mainland

Some of population A is carried from the mainland to the island.

2

A   B

Isolated from population A, the island population eventually evolves to form a new species, B.

3

A   B

Some of population B moves over the mountains to the opposite side of the island.

4

drier warmer climate   wetter colder climate

A   B   C

Isolated from population B, a new species, C, evolves due to the different selection pressures on that side of the mountain range.

There are a number of other reasons, in addition to geographical isolation, why two species may not be able to breed and produce fertile offspring. These are shown in Table 5.4.

Table 5.4 Reasons why two species may not be able to interbreed successfully in addition to geographical isolation.

Method of isolation	Description
ecological isolation	The species occupy different parts of the habitat. For example, the violet species *Viola arvensis* grows on alkaline soils, whereas *V. tricolor* grows only on more acidic soils (Figure 5.51).
temporal isolation	The species exist in the same area but reproduce at different times.
behavioural isolation	The species exist in the same area, but do not respond to each other's courtship behaviour (e.g. different species of firefly).
physical incompatibility	Species coexist, but there are physical reasons that prevent them from copulating (e.g. size or shape of genitals in some insects).
hybrid inviability	In some species, hybrids are produced but they do not survive long enough to breed.
hybrid sterility	Hybrids survive to reproductive age but cannot reproduce (e.g. mules are a cross between donkeys and horses; they are very hardy but cannot reproduce).

A

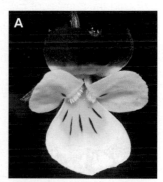

B

Figure 5.51 The violet species *Viola arvensis* (**A**) and *Viola tricolor* (**B**) are reproductively isolated because they usually grow on different soil types.

It is not always clear cut!

Ruddy ducks and white-headed ducks (Figure 5.52) are accepted as two different species, but when ruddy ducks were introduced from the US to Europe they interbred to produce fertile hybrids. The white-headed duck is now an endangered species. Under the Convention on Biological Diversity, the UK is obliged to conserve the white-headed duck. The UK Government is introducing measures to protect the white-headed duck, including attempting to eradicate the UK ruddy duck population.

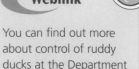

Weblink

You can find out more about control of ruddy ducks at the Department for Environment, Food and Rural Affairs website.

🔺 **Figure 5.52 A** White-headed duck. **B** Ruddy duck. These ducks look very different but are they actually the same species?

Q5.28 **a** Should ruddy ducks and white-headed ducks be considered to be one species or two?
b Should the number of ruddy ducks be controlled?

Did you know? Adaptive radiation and convergent evolution

Natural selection produces organisms that are better adapted to their environments. This fact makes the lives of taxonomists much harder, because it leads to two differing phenomena.

◀ **Figure 5.53** A dolphin (top left), a shark (middle) and an ichthyosaur (bottom) look similar as a result not of common ancestors but of similar selective pressures.

The first of these is called **convergent evolution**: for example, animals that are unrelated, but live in similar habitats and occupy similar niches, experience the same selective pressures, and often end up looking very similar. For instance, dolphins, sharks and ichthyosaurs, shown in Figure 5.53, all share a similar shape which is adapted for fast swimming and catching medium-sized fish, but belong to different classes of vertebrates (mammals, fish and reptiles).

The second phenomenon is called **adaptive radiation**. Closely related species may look very different, as a result of having adapted to widely differing niches. For example, the phylum Mollusca includes bivalves such as giant clams adapted to filter feeding, gastropods such as grazing snails and sea slugs, and octopuses and squids which are predators capable of considerable bursts of speed. All these are believed to have evolved from a worm-like ancestor, but are the result of selection in very different niches.

Why is the theory of evolution controversial?

Darwin had considered training to be a clergyman and knew that his ideas about evolution meant that he seemed to be challenging the Bible. The first two chapters of Genesis record how all of the Earth's organisms were created by God about 6000 to 10 000 years ago, in much the form that they exist today.

It is hard for many people to appreciate just how controversial Darwin's ideas were at the time. To this day there are many people in the world who cannot reconcile their religious beliefs and the theory of evolution. Few Muslims accept it and many Christians either reject or aren't comfortable with it. In the UK around 10% of people are *creationists*. Creationists accept the literal teaching of the Bible, Qur'an or other scriptures and reject the theory of evolution. Creationists believe, for example, that God specially created Adam and Eve out of the dust of the earth so that humans are quite distinct from all other species. In the USA around 40% of people are creationists. Indeed, worldwide there is no doubt that the idea that all organisms are descended from a common ancestor that lived some 3000 million years ago – which is what evolutionary biologists believe – is a minority position.

The great majority of biologists, geologists and other scientists do accept a modernised version of evolution called *neo-Darwinism*. Neo-Darwinism combines Darwin's and Wallace's theory of natural selection with what we now know about inheritance. Many scientists who accept the theory of evolution in this modernised version also have firm religious beliefs. Such scientists have various ways of reconciling their religious beliefs with the theory of evolution. For example, they may believe that God created the original conditions that enabled the universe to come into existence, and then allowed evolution to take its course. In this understanding, God gives the whole of creation, including us, a certain freedom. Life is not predetermined but open ended.

> **Activity**
>
> Discuss why the theory of evolution is controversial using the role play in **Activity 5.16. A5.16S**

5.7 On the edge of extinction

Extinction on the increase

There are growing concerns about the number of species that are threatened with extinction, and that biodiversity is facing a crisis. Why should we be concerned? Species have been going extinct since life first evolved on Earth over 3500 million years ago. Darwin brought back the remains of many extinct species from his expeditions on the *Beagle*. However, the current species extinction rate is estimated to be between 1000 and 10 000 times higher than it would naturally be.

Perhaps the most famous of an anthropogenic extinction was the dodo (*Raphus cucullatus*), a flightless bird from the island of Mauritius in the Indian Ocean that died out in 1681 (Figure 5.54). The Portuguese sailors who landed on the island back in 1600 named this species the dodo due to its apparently docile and simple nature. What we know now is that being an island species it was not adapted to deal with predators, as there were no native predators on the island. Some dodos were eaten by the sailors, but the principal threat to the species was the destruction of its natural forest habitat and the introduction of rats, pigs and cats, predators that destroyed the dodos' nests and ate their young chicks.

Since the time of the dodo, over 800 named species are known to have become extinct. This is undoubtedly a great underestimate of the actual number of species lost. IUCN, the World Conservation Union, identifies over 11 000 species known to be threatened with extinction.

▲ **Figure 5.54** All that remains of the dodo are illustrations, a few written accounts, some skeletons and models made of chicken feathers.

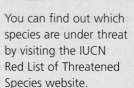

Weblink

You can find out which species are under threat by visiting the IUCN Red List of Threatened Species website.

Conservation of species

There are several arguments for conserving species. Here we discuss four – in no particular order!

Economic arguments

Species are valuable to humans in a number of ways, the most obvious use being food. As we saw in Topic 4, we also obtain a seemingly endless list of industrial products from the living world – such as wood, fibres, gums and dyes – and there are potentially many more natural products out there. We use both plant and animal species for sourcing medicines. For example, leeches produce anticoagulants used in surgery, and willow trees were the original source of aspirin. Scientists have estimated that the value of the undiscovered drugs from tropical rainforests is around $150 billion. So the rainforests may be worth far more to us alive than they are dead. Reserves, zoos and wildlife farms can be profitable: wildlife tourism plays an important role in many countries' economies.

Figure 5.55 Poison or potion? The golden poison dart frog (*Phyllobates terribilis*). The toxin on this frog's skin contains chemicals which can be used as human heart stimulants – one good reason to conserve this species.

Living organisms also provide a bank of genes for use in the breeding of new crops, whether by traditional methods or genetic manipulation.

Aesthetic reasons

Many of us enjoy walking in the countryside or relaxing in parks. The wildlife we see is an important part of that experience, whether we are watching a squirrel in the park or seeing 'big game' on an African safari. Every species is unique and irreplaceable, having its own particular beauty (like a work of art). Once extinct, it can never be recreated. Humans should conserve species and their habitats because they enrich our lives.

Ethical arguments

It can be argued that all species have a right to exist and that we have a duty not to allow any species to go extinct at our hands. Humans are just one of millions of species on Earth, but with the power to control the destinies of other species. This power should be used responsibly. A culture that encourages respect for wildlife is preferable to one that does not.

Ecological concerns

No species exists in isolation; all are interdependent in ways that we still do not fully understand. The loss of one species has consequences for many others. As we saw earlier, in a tropical rainforest there may be in excess of 160 insect species that are unique to each tree species. Loss of the tree species means the loss of all these insects and any other species that rely on them or the tree. In addition, trees play a vital role in maintaining the soil. Removing trees can lead to soil erosion, flooding and drought. Extinction of a species is a symptom of deterioration of the natural environment, and tends to produce ecological instability. Any instability ultimately affects humans, since the Earth is our environment too. Stability is essential in the long term for sustained food production.

◀ **Figure 5.56** The golden lion tamarin (*Leontopithecus rosalia*) was on the edge of extinction in the early 1970s with only about 200 animals left in the wild. In 1974, the Golden Lion Tamarin Conservation Programme was set up. Scientists, conservationists and educators have worked together to protect and study the tamarin and its habitat so as to overcome the continuing threats to the species. Habitat management, captive breeding, reintroductions to the wild, education and research are all part of this international rescue effort to save the species. The golden lion tamarin is a conservation success story. Assessed as Critically Endangered in 2000 it has now been downgraded to Endangered as a result of nearly 30 years of conservation efforts. There are now about 1000 golden lion tamarins in the wild, with a further 500 in zoos involved in the conservation programme across the world.

Conservation methods

The primary threat to most species and habitats is human activity. This includes land development (which causes habitat destruction, fragmentation and degradation), over-exploitation, introduction of alien species, and pollution.

The growing number of threatened species will not be saved from extinction merely by putting a fence around their remaining habitat and hoping for the best. As the golden lion tamarin story (Figure 5.56) illustrates, conservation requires a coordinated approach. Ideally, conservation management should be applied on site (*in situ*), protecting ecosystems and maintaining fragile habitats. In addition, vulnerable populations in the wild can be supported with captive breeding populations in zoos if genetic diversity is maintained. This is known as off-site (*ex situ*) conservation.

It is most important to remember that large, attractive species such as the golden lion tamarin will always attract attention and conservation effort, but less 'attractive' species – perhaps slugs, mosses or biting insects – also deserve attention. The conservation of flagship species, like tigers and pandas, attracts public attention. The management strategies which can then be implemented will help protect habitat which can then support other, less 'attention-grabbing' species.

Activity

Activity 5.17 allows you to examine the threats to golden lion tamarins and how they and their habitat are being conserved. **A5.17S**

5.8 Protecting ecosystems

Table 5.5 shows threats to species and habitats in the UK. Notice how habitat destruction is the major threat. Similar problems are faced by ecosystems worldwide – destruction of natural ecosystems is the greatest threat to biodiversity.

Threat	Number of habitats affected (n = 45)	Number of species affected (n = 391)
habitat destruction	34	164
habitat fragmentation	17	34
change in agricultural management	15	49
agricultural intensification	29	116
lack of appropriate management	21	137
afforestation	13	28
changes to woodland management	5	22
water management	18	90
coastal development and management	23	8
water pollution	28	70
air pollution	13	14
climate change/sea level rise	21	22
recreational pressure	16	39
fisheries management	19	34
accidental killing		23
collecting/taking	—	25
invasive or introduced species	—	26
predation/disease	—	23

 Table 5.5 Threats to 45 habitats and 391 species identified as priorities for conservation in the UK Biodiversity Action Plan.

Weblink

To find out more about SSSIs, NNRs and Marine SACs (Special Areas of Conservation) visit the English Nature website.

Most of the habitats referred to in Table 5.5 are designated as National Nature Reserves (NNRs) and Sites of Special Scientific Interest (SSSIs). The aim of identifying these areas is both to protect them and to ensure that there is appropriate management of the sites.

If an ecosystem is stable it will not require any intervention beyond avoiding change of use. However, many ecosystems that we want to conserve are not stable. They are dynamic, and must be managed if their biodiversity is to be maintained.

Communities change

When Darwin arrived on the Galapagos Islands he was amazed by the diversity of organisms. But it had not always been as he found it. The islands are volcanic in origin, and there are still active volcanoes on some of them. Globally, it is not uncommon for new islands to be created from the lava expelled from an active volcano. Surtsey was formed in this way in 1963 off the coast of Iceland. The bare rock and ash of such an island does not stay bare for long. Plant or lichen species which can cope with the harsh conditions on the bare rock become established. This creates a community which soon changes over time, in a process called **succession**.

 **Figure 5.57** Lichens on a rock surface can start the process of breaking it up to form soil.

Primary succession

A **primary succession** starts in newly formed habitats where there has never been a community before. This may occur on bare rock, on material on the seashore like sand and shingle, and in open water, for example. Unless prevented, succession then continues until, sometimes thousands of years later, a relatively stable community is established.

The pioneer phase: enter the botanical mavericks

The first organisms to colonise bare rock are lichens and algae, **pioneer species**. These are the only species that can cope with the extremes of temperature and lack of soil, water and nutrients (Figure 5.57).

The pioneers start to break up the rock surface, allowing some organic material to accumulate with the broken-up rock as the beginnings of soil. In so doing, they change the conditions in the habitat just enough to make them suitable for other species. Wind-blown moss spores start growing.

Succession continues

The mosses build up more organic matter in the soil which can then hold water. The development of a soil enables seeds of small shallow-rooted plants to establish.

△ **Figure 5.59** Once plants become established animals, like this dune snail, can also colonise.

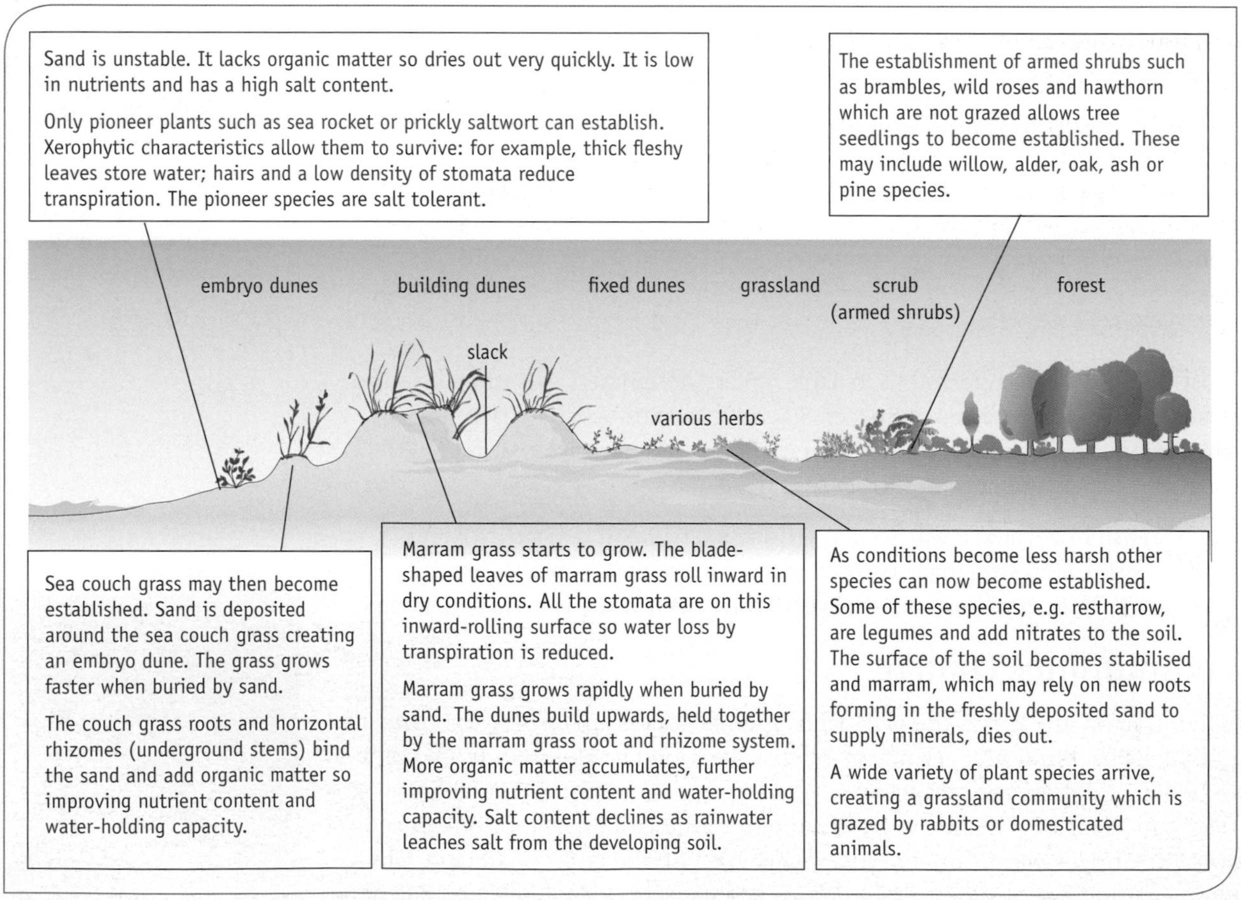

Sand is unstable. It lacks organic matter so dries out very quickly. It is low in nutrients and has a high salt content.

Only pioneer plants such as sea rocket or prickly saltwort can establish. Xerophytic characteristics allow them to survive: for example, thick fleshy leaves store water; hairs and a low density of stomata reduce transpiration. The pioneer species are salt tolerant.

The establishment of armed shrubs such as brambles, wild roses and hawthorn which are not grazed allows tree seedlings to become established. These may include willow, alder, oak, ash or pine species.

embryo dunes   building dunes   fixed dunes   grassland   scrub (armed shrubs)   forest

slack

various herbs

Sea couch grass may then become established. Sand is deposited around the sea couch grass creating an embryo dune. The grass grows faster when buried by sand.

The couch grass roots and horizontal rhizomes (underground stems) bind the sand and add organic matter so improving nutrient content and water-holding capacity.

Marram grass starts to grow. The blade-shaped leaves of marram grass roll inward in dry conditions. All the stomata are on this inward-rolling surface so water loss by transpiration is reduced.

Marram grass grows rapidly when buried by sand. The dunes build upwards, held together by the marram grass root and rhizome system. More organic matter accumulates, further improving nutrient content and water-holding capacity. Salt content declines as rainwater leaches salt from the developing soil.

As conditions become less harsh other species can now become established. Some of these species, e.g. restharrow, are legumes and add nitrates to the soil. The surface of the soil becomes stabilised and marram, which may rely on new roots forming in the freshly deposited sand to supply minerals, dies out.

A wide variety of plant species arrive, creating a grassland community which is grazed by rabbits or domesticated animals.

△ **Figure 5.58** Sand-dune succession. The land now covered by forest was originally the sandy foreshore and has gone through all the stages described to become the climax community. The succession literally pushes the sea back and turns a seashore into a forest.

As the conditions in the habitat improve, seeds from larger, taller plants appear. They compete with the plants already present in the habitat and, winning the competition, they replace the existing community.

Eventually a community usually dominated by trees is reached, and this stable **climax community** often remains unchanged unless conditions in the habitat change. The nature of the climax community depends very much on the environmental conditions, such as climate, the soil, and which species are available. In much of Britain below 500 m the natural climax is forest; this is oak dominated in many areas, but with beech or ash on limestone and with birch on more acidic soil.

The dominant species of a community is the one that exerts an overriding influence over the rest of the plant, microbe and animal species. (Sometimes several species will share the role of being dominant and are said to be codominant.) The dominant species is usually the largest and/or most abundant plant species in the community, and this often gives the community its name, for example, an oak forest.

As the succession progresses, the number of niches increases, as does the number of species present. However, it is not unusual for the climax community to have lower biodiversity than the preceding stages in the succession. Figure 5.58 shows the stages in a typical sand-dune succession.

> **Activity**
>
> Use the interactive tutorial in **Activity 5.18** to follow succession in one habitat. **A5.18S**

A climax community without trees

There are some habitats where the climax community is not a forest or woodland. The large areas of bare stony soil on a hill in Shetland called the Keen of Hamar (Figure 5.60) looks like an abandoned motorway construction site which will soon grass over as succession gets underway. But succession does not continue, apparently because of the properties of the soils and underlying rock. One possibility is that the soil is so sandy and freely draining that it suffers drought conditions after only a few days without rain. The vegetation resembles a low-competition pioneer phase of a more typical succession yet is, in this case, the climax community and probably has been for over 10 000 years.

The Keen of Hamar serpentine debris habitat has high biodiversity including rare species and is the best example of its kind in North-west Europe. The distinctive species are slow growing and could not compete successfully with vigorously growing grasses. Their survival depends on the habitat remaining an open pioneer habitat.

▲ **Figure 5.60 A** Keen of Hamar National Nature Reserve, Unst, the Shetland Islands. Historical data and detailed recording of permanent quadrats on serpentine debris (**B**) show that succession is not happening.

Secondary succession

On bare soil where an existing community has been cleared, **secondary succession** occurs. When natural plants start to grow in a ploughed field or after a forest fire they mark an early stage in secondary succession. In the absence of human interference, secondary succession would, in many parts of the UK, lead to the re-establishment of a forest climax community. In the same way the lost Inca cities were swallowed up by the South American rainforest.

Bare soil does not stay bare for long in nature. Seeds of many species will already be lying dormant in the soil (as a seed bank), and others will be brought by the wind or animals. Groundsel (*Senecio vulgaris*) is an example of a pioneer species adapted to take advantage of newly bare soil where there is little or no competition (Figure 5.61). These adaptations include:

- seeds widely dispersed by the wind
- rapid growth rate
- short life cycle
- abundant seed production.

Pioneer species like groundsel cannot compete with slower-growing species such as grasses. Groundsel succeeds by getting in quickly (effective, widespread dispersal), growing rapidly and flowering within a few weeks.

Deflected succession

A community that remains stable only because human activity prevents succession from running its course is called a **deflected succession** (Figure 5.62). For example, sheep grazing in Britain prevents many grasslands from developing into woodland. When, one day, you come out of your house and notice a wood where once there was a garden you have really put off mowing the lawn too long this time!

Many habitats such as chalk grassland of the South Downs or heather moors need to be actively managed to prevent succession which would result in a loss or change in biodiversity. Grazing, mowing or burning may be used to deflect succession.

▲ **Figure 5.61** Groundsel: a common weed that is a pioneer species.

> **Checkpoint** ✓
>
> **5.6** Draw up a table of comparison for primary and secondary succession and explain what is meant by deflected succession.

◀ **Figure 5.62** A golf course or grazing on the grassland behind sand dunes deflects succession, preventing the formation of a climax community.

5.9 Off-site (*ex situ*) conservation

The role of zoos

Figure 5.63 London's Royal Menagerie, 1812.

The history of present-day zoos began in the 1750s, with the foundation of the first zoo in Europe by Emperor Franz Stephan in the grounds of Schönbrunn Palace just outside Vienna. London Zoo was founded in 1826, some six years before Charles Darwin set sail in the *Beagle*. Initially the collection was privately studied by eminent scientists of the day, but it opened its doors to the public in 1847. In those days before cameras and television, the public flocked to zoos, circuses and sideshows to witness the wonders of the natural world. But almost nothing was known of these exotic animals' biology or behaviour, and little attention was paid to their needs in captivity. They were usually kept alone in bare cages designed for maximum visibility by the public (Figure 5.63).

Today, zoos attract over 600 million visitors each year throughout the world. They manage over 1 million vertebrates in over 1000 registered collections and aim to have a significant impact in conservation, research and education.

Centres for scientific research

One argument put forward for zoos is that they can play a vital role as research centres, enabling us to understand how to conserve particular species. The example below of mountain chickens at Jersey Zoo illustrates this.

How do mountain chickens breed?

The mountain chicken (*Leptodactylus fallax*) is not, as its name suggests, a bird, but is in fact one of the largest frogs in the world (Figure 5.65). It is found only on the Caribbean islands of Dominica and Montserrat. The mountain chicken's numbers have been declining because it has long been a national dish of the islands. It is hunted in large numbers for its meaty legs, which are used in traditional West Indian recipes. As their name suggests, their taste is somewhat like that of chicken.

Did you know? Questioning the role of zoos

There is a growing awareness among those who run zoos of issues surrounding animal welfare, and the role zoos can play in saving species from extinction (Figure 5.64). As we learn more about habitat protection and the need to conserve endangered species in their natural environments, the conservation role of zoos and the justification for keeping wild animals in captivity are being questioned.

Some endangered species are extinct in the wild and exist only in zoos. Few people argue that they would be better off dead than living in captivity. Most zoos now try to have their animals living in naturalistic environments. UK zoos are regulated by the Zoo Licensing Act (1981), which aims to promote minimum standards of welfare, meaningful education, effective conservation, valuable research and essential public safety. The introduction of the European Zoos Directive (1999/22/EC) by the European Union places a greater emphasis on conservation, education, research and welfare.

In 1964, Bill Travers and Virginia McKenna starred in the film *Born Free*, which told the true and very moving story of the return of Elsa the lioness to the wild. The Born Free Foundation is now an international wildlife charity working to prevent animal cruelty, alleviate suffering and encourage everyone to treat all individual animals with respect.

▲ **Figure 5.64** Keeping elephants in captivity – should zoos take into account that in the wild most elephants live in family groups and range over very large areas?

The Born Free Foundation believes wildlife belongs in the wild. It is dedicated to the conservation of rare species in their natural habitat, and the phasing out of traditional zoos. In 2000 it undertook a survey of the health status of UK zoos and published the report Zoo Health Check 2000.

The Born Free Foundation report expresses concerns about the welfare of many animals in zoos, having observed many animals exhibiting stereotypic behaviours such as repetitive pacing up and down and chewing of bars. They found that up to one in five collections that qualify as zoos under the current legislation do not have a zoo licence. Perhaps more importantly from the animals' perspective, 95% of species in zoos are not endangered, and are not part of European captive breeding programmes. The Born Free Foundation would rather see animals being bred in protected habitats in the wild.

Q5.29 What arguments might be presented for and against keeping animals captive in zoos?

Extension

Extension 5.8 allows you to consider some differing views on the role of zoos and debate the future of zoos.
X5.08S

Since 1995 over 75% of Montserrat has been engulfed by the continuing activity of the Soufrière Hills volcano. This volcano has destroyed much of the frog's remaining rainforest habitat and left the already threatened species in a very precarious situation.

▲ **Figure 5.65** *Leptodactylus fallax* – the frog that's a chicken. These frogs grow to 21 cm in length.

▲ **Figure 5.66** Unlike other frogs, mountain chickens lay their eggs in underground nests, covering them in foam.

Little was known about the reproductive biology of *L. fallax*. The main reason for this was that it leads a secretive nocturnal life and so has proved difficult to study in the wild. If it was to be saved from extinction, more information was needed about its needs. In 1999, 13 frogs were taken to Jersey Zoo, allowing keeper scientists to find out more about their elusive lifestyle. Breeding trials have been set up to establish a captive breeding programme.

Q5.30 **a** Suggest why the mountain chicken's underground nest-building behaviour (Figure 5.66) has evolved.
b What would be some of the complications of conducting a study on these frogs in the wild?

Captive breeding programmes

Until recently, animal collections were unplanned and opportunistic. Zoos simply bought their animals from explorers and traders who would return from expeditions in ships laden with crates of anything they could catch and manage to bring back alive.

Today an important role of zoos is the successful breeding of the animals in their care. The aims of a captive breeding programme include:
- increasing the number of individuals of the species if numbers are very low
- maintaining genetic diversity within the captive population
- reintroducing animals to the wild if possible.

There are European Endangered Species Programmes for over 200 species, with more than 400 institutions in Europe participating. Each species has an appointed coordinator/studbook-holder who advises which animals should or should not breed and on the movement of animals between partner zoos, in order to maximise genetic diversity within the captive populations.

Weblink

To find out more about captive breeding programmes visit the European Association of Zoos and Aquaria website.

Genetic uniformity, where individuals within a population have similar genotypes, can be an advantage in a stable environment. But if the environment changes, a new disease emerges or a population moves, a genetically diverse population will be at an advantage.

Scientists interested in the preservation of threatened species hope that there is sufficient genetic diversity in the population to ensure that some of the individuals can cope with the new conditions so the population survives. This is natural selection and results in adaptation, the accumulation of genotypes favoured by the environment. No natural environment remains unchanged forever. It is therefore essential for long-term survival that populations should be able to evolve as a result of natural selection. Such evolution is unlikely unless there is genetic variation for evolution to act upon.

How genetic variation is lost

Genetic drift

In a small population, some of the alleles may purely by chance not get passed on to offspring, as shown in Figure 5.67. This change in the allele frequencies over time is known as **genetic drift**, and leads to a reduction in genetic variation.

Inbreeding depression

In a small population, whether in the wild or in captivity, the likelihood of closely related individuals mating increases. This inbreeding causes the frequency of homozygous genotypes to rise, with the loss of heterozygotes. Inbreeding results in individuals inheriting recessive alleles from both parents, and the accumulation of the homozygous recessive genotypes in the offspring. Many recessive alleles have harmful effects so **inbreeding depression** results. The offspring are less fit (less able to survive and reproduce). They may be smaller and not live as long, and females may produce fewer eggs.

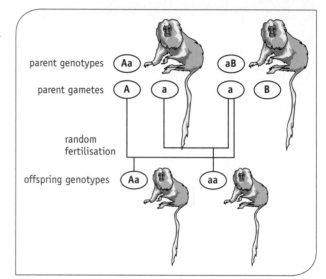

▲ **Figure 5.67** Genetic drift leads to less variation. In the example shown, the parents carry three alleles for a particular characteristic; by chance one of the alleles is not passed on.

Conserving genetic diversity

Conservation *in situ* (on site) to maintain the size of wild populations is the best way to prevent genetic drift and inbreeding depression. However, *ex situ* (off-site) conservation can also play a role.

Keeping studbooks

The studbook for an individual species shows the history and location of all the captive animals of that species in the places that are cooperating in an overall breeding plan. London Zoo, for example, keeps studbooks for many species including the slow loris, the Sumatran tiger (Figure 5.68) and the ocelot, and contributes to the studbook data for other species such as the golden lion tamarin.

Extension

Discover some of the disadvantages and advantages of genetic uniformity in the ant supercolony story in **Extension 5.9**. **X5.09S**

Studbooks provide the raw data upon which all the breeding plans are based. The scientists' understanding of genetics shapes the breeding plans themselves. Conservation scientists must ensure that genes from all the founder members of the population (the original group of individuals, usually wild caught, on which the current population is based), or at least all remaining breeding adults, are retained and are *equally* represented in the subsequent generations (assuming that these founders are unrelated). This requires that individuals who breed poorly in captivity must be encouraged to breed, whilst those that are particularly good breeders must be limited in their breeding success.

▲ **Figure 5.68** The Sumatran tiger is an endangered species which has been supported by a captive breeding programme in zoos. Zoo managers make decisions about breeding these animals based on their studbook data.

This approach, using a breeding plan, is a very different principle from that followed by zoos of 50 years ago, when they simply raised their captive populations from the best breeders. This seemed like common sense, but it obviously reduced genetic variation and began a process which made it less likely that there could be a successful reintroduction to the wild.

Activity

Investigate the animal dating agency for lemurs in **Activity 5.19**. **A5.19S**

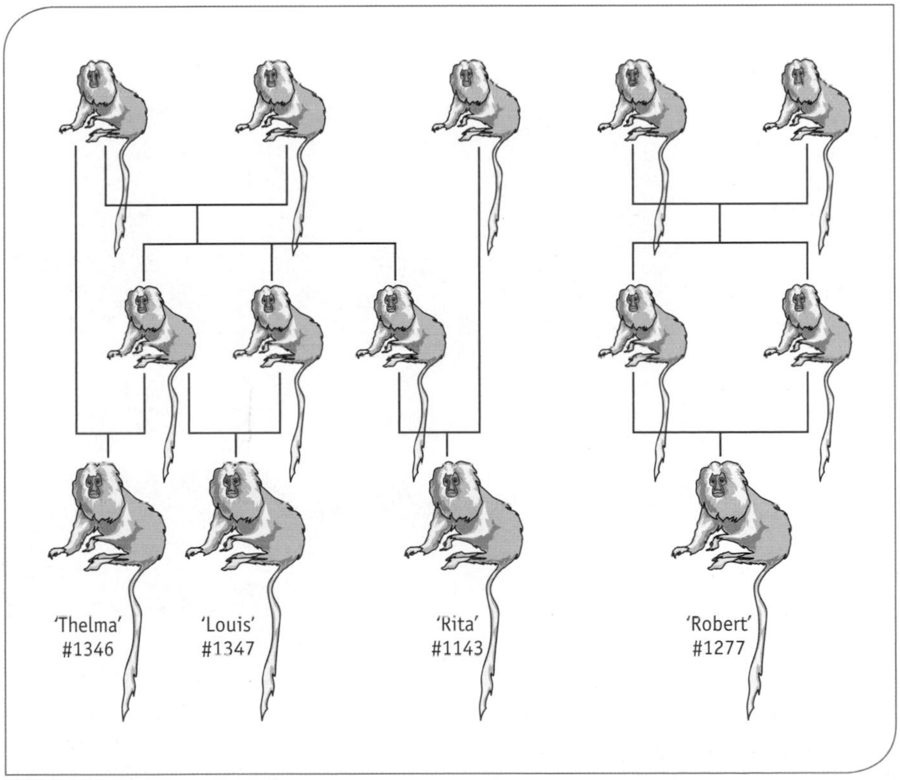

'Thelma'
#1346

'Louis'
#1347

'Rita'
#1143

'Robert'
#1277

◀ **Figure 5.69** Some of the breeding undertaken as part of the golden lion tamarin programme.

Q5.31 Suggest which of the named animals in Figure 5.69 is genetically most valuable for breeding purposes.

More and more, studbook records are being supported by techniques of cytogenetics – looking at the structure of chromosomes – and of molecular biology – studying the nature of the genes themselves. These additional techniques are important because studbook data may be incomplete (zoos in the past did not keep records as carefully as they do now). In addition, these new techniques can reveal whether some individuals are more closely related than is desirable for breeding purposes.

Reintroducing animals to the wild

One role that zoos have in species conservation is captive breeding for reintroduction. This involves breeding animals in captivity that are then returned to their native habitats. Clearly this can only work if the habitat is still intact! Reintroduction is a complicated process, particularly when species need to learn new skills before they have the ability to survive in the wild. However, there are cases in which zoos have successfully reintroduced captive-bred animals to the wild, as the example below illustrates.

Going, going, ... saved!

The Mauritius kestrel (*Falco punctatus*) was one of the rarest birds in the world, consisting of just four known individuals in 1974. The population was so small partly due to the extensive use of DDT as a pesticide to kill malaria-carrying mosquitoes. As the chemical worked its way up the food chain and accumulated, it was the kestrels as top predators that were most affected. Furthermore, the native habitat had become so degraded due to the removal of hardwood trees that it was not able to support a rapid increase in kestrel numbers without additional management techniques.

Carl Jones went out to Mauritius soon after 1974 and realised that the species could be saved. With the backing of the Durrell Wildlife Conservation Trust whose headquarters are at Jersey Zoo he set up a captive breeding centre on Mauritius, taking eggs from wild birds and hatching them in captivity (Figure 5.70). The young birds were then either returned to wild pairs for rearing or were 'hacked out'. This meant taking them out into the forest and gradually giving them more freedom and less food to encourage them to feed for themselves. Nest boxes were provided and natural cavities in trees modified to increase the number of suitable nest sites. In addition, predators, especially rats, were controlled at release and nest sites. As a result, there are now over 800 Mauritius kestrels slicing through the forest air.

▲ **Figure 5.70** Fledgling captive-bred Mauritius kestrels *Falco punctatus*.

Activity

In **Activity 5.20** you analyse the reintroduction programmes for the Mauritius kestrel and ruffed lemurs. **A5.20S**

Checkpoint

5.7 Summarise the roles of zoos in the conservation of endangered species.

Q5.32 What would be both the advantages and disadvantages of keeping young Mauritius kestrels in an open nest box in the forest?

5.10 The human dimension

Conserving wildlife and resources

Exploiting wildlife

We humans currently use 10–20% of the Earth's primary productivity. This creates a conflict between humans and wildlife. Wildlife is valuable to us – many species and habitats are used as resources both in the UK and elsewhere across the globe. Woodlands provide timber, wetlands are used for waste-water treatments, floodplains and coastal dunes provide natural erosion control, and fish and other wild species are important sources of food and medicines.

Wildlife is viewed by many as a freely exploitable resource. Millions of wild animals and plants, worth many billions of pounds, are traded every year, although a significant amount of the trade is illegal. Human activity such as over-exploitation, habitat destruction and pollution threatens these wildlife resources.

The killing of huge numbers of animals for bush meat (Figure 5.71), much of it illegal, to meet basic food needs and as a source of income is causing a rapid decline in wild population sizes of thousands of species. It has been estimated that between 70 000 and 170 000 square kilometres of tropical rainforest are lost annually (equal to between 21 and 50 soccer fields per minute). Much of the logging is illegal. Commercial logging and clearing land for development and agricultural use destroys or fragments forests, putting the species they contain at risk. This threatens the survival of numerous species, upsets the delicate balance of ecosystems and destroys valuable resources.

Once you put human needs into the conservation equation, simple preservation of a species or habitat, by cutting it off from the encroaching world, is rarely an option. It is likely that a range of strategies will be required to protect wildlife species and their habitats, such as the development and enforcement of legislation to protect wildlife, and sustainable management of wildlife resources. The involvement of local people in managing and protecting their resources is supported by organisations like TRAFFIC, the joint wildlife monitoring programme of the World Wide Fund for Nature (WWF) and IUCN The World Conservation Union.

Can legislation help?

Several international conventions have been agreed which aim to conserve wildlife and their habitats. In 1946 the International Convention for the Regulation of Whaling was signed in Washington. The objective of the Convention was to conserve whale stocks while providing for the orderly development of the whaling industry. The International Whaling Commission (IWC) is responsible for implementation of this Convention. Quotas were set in an attempt to make whaling sustainable. However, during the 1960s and 1970s the numbers of whales caught far exceeded the quotas. For example, it is estimated that between 1949 and 1980 the USSR catch included 3200 right whales, yet these had been completely protected on conservation grounds since 1935. (Right whales are so named because they are the easiest whales to catch. Their numbers were therefore the first to decrease as a result of human activity.)

▲ **Figure 5.71** These animals were killed for food as 'bush meat'. Over-exploitation of wild animals in this way can threaten species survival.

Activities

Activity 5.21 'The good, the bad and the ugly' lets you brainstorm some ideas about the effects of human populations on wildlife. **A5.21S**

In Activity 5.22 you can examine some of the conflict between wildlife and humans around Lac Alaotra, the largest freshwater lake on Madagascar, and discuss how the problems might be overcome. **A5.22S**

At the IWC meeting in June 1982, growing international concern for the survival of whales prompted the Commission to issue a moratorium (temporary ban) on commercial whaling of all whale stocks from 1985/86. The moratorium did not affect aboriginal subsistence whaling which is permitted from Denmark (Greenland: fin and minke whales), the Russian Federation (Siberia: gray and bowhead whales), St Vincent and the Grenadines (humpback whales) and the USA (Alaska: bowhead and occasionally gray whales). In addition, certain countries, including Japan, have continued to catch whales for so-called 'scientific research purposes'. Meat from whales caught in this way is allowed to be sold as food.

Whales are also protected under the Convention on International Trade in Endangered Species (CITES). Signed by over 90 countries in 1973, CITES attempts to control the international trade in wildlife and wildlife products. Imports, exports and re-exports have to be authorised through a licensing system. The species covered by CITES are listed in one of three categories:

- Appendix I species are threatened with extinction, and trade in these species is effectively banned.
- Appendix II species are not necessarily threatened with extinction but trade in them must be controlled to ensure that their survival does not become threatened.
- Appendix III species are protected in at least one country, which has asked other countries for assistance in controlling the trade.

Cultural differences between peoples can mean that nations have differing views with regard to how they value wildlife. Where one country may consider that the conservation of a threatened species requires an outright ban on any trade, others see wildlife as resources that can be utilised as food, clothing, timber, leather goods, tourist curios and medicines.

Conserving biodiversity and sustainable development

Q5.33 **a** From the graph in Figure 5.72, in 2005 the human population was about 6500 million (6.5 billion). It is estimated that by 2050 it will be 9000 million. Approximately what percentage of the world's primary productivity might we be using in 2050? (Current use is approximately 10–20%.)
b The rate of resource use has increased faster than the population size. Suggest why this might be the case.

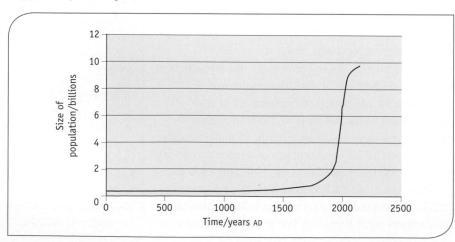

◀ **Figure 5.72** Graph of world human population size as a function of time. It has been calculated that it would take four more planets to enable the current world human population to have the same standard of living as those now living in the USA.

In 1992 the United Nations Conference on Environment and Development was held in Rio de Janeiro, Brazil. Often referred to as the Rio Summit, this brought 159 heads of state and government together to debate the essential link between humanity and the Earth. The outcome was a blueprint for sustainable development in the twenty-first century called Agenda 21. A sustainable development is one that meets the needs of the present without compromising the ability of future generations to meet their own needs. It was the first time that so many world leaders had formally recognised that every single one of us has a right to live our life, but not at the expense of the environment in which we live.

We all have an impact on the world around us, which can be either positive or detrimental. This action may be direct, through planting a tree or cutting it down, or indirect, such as recycling cans or buying products grown on land cleared of its native forest for the crops. If we are to achieve a sustainable level of use of natural resources then action needs to be taken at individual level as well as at local level, nationally and globally.

At the Rio Summit, nations also agreed the Convention on Biological Diversity, although all nations have not yet signed the agreement. The objectives of this convention are the conservation of biological diversity, the sustainable use of its components, and the fair and equitable sharing of the benefits arising from the utilisation of genetic resources.

Each country that signs the Convention has to identify and monitor the components of biological diversity important for conservation and sustainable use. They must also develop national strategies, plans or programmes for the conservation and sustainable use of biological diversity.

The Convention on Biological Diversity has been incorporated into UK law and, in 1994, the UK Government published their strategy: The UK Biodiversity Action Plan (UK BAP). This aims to ensure that there is no further net loss of biodiversity and some restoration of past losses. Under the terms of the UK BAP, 391 Species Action Plans and 45 Habitat Action Plans have been published. Each plan includes a statement of the current status of the species or habitat, factors causing loss or decline, conservation objectives and targets, and proposed actions. Not only are there National Action Plans but Local Biodiversity Action Plans are also being developed. These are encouraging signs but the next ten or so years will tell whether we have finally begun in the UK to reverse literally thousands of years of habitat and species loss.

Activity

Use **Activity 5.23** to check your notes using the topic summary provided. **A5.23S**

Topic test

Now that you have finished Topic 5, complete the end-of-topic test before starting Topic 6.

Topic (6) Infection, immunity and forensics

Why a topic called Infection, immunity and forensics?

Two deaths, two bodies discovered. There is a mystery: who were the dead man and woman? Forensic biologists and pathologists use a wide range of techniques to help answer such questions, to determine when people died and what caused their untimely deaths.

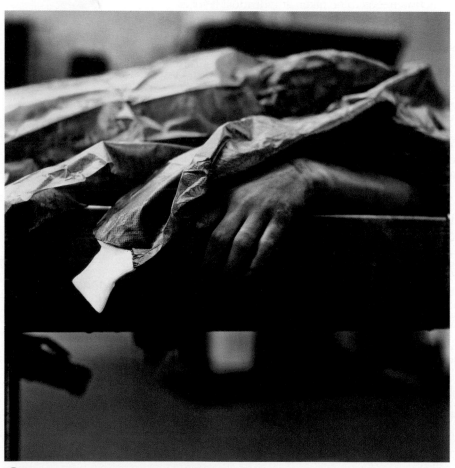

△ **Figure 6.1** Whose body? When and how did they die?

The villains in this particular case were not violent criminals or gun-carrying muggers who made an obvious assault. The victims may have died because of the genes they inherited, the food they ate or other aspects of their lifestyle. Microbes may have stolen into their bodies undetected. But if a disease-causing pathogen enters your body, it may soon make itself at home in which case you will know all about it.

Sometimes an invading microbe will be nothing more than a nuisance, causing a cold or sore throat. The immune system swings into action and deals with the uninvited guests. Specialised cells and chemicals, functioning like a well organised army, destroy the invaders. But sometimes the microbes are so numerous that without help the immune system is overwhelmed and the body faces disaster. Did this happen to either of our victims?

Antibiotics, discovered in a lucky accident by Alexander Fleming in 1928, revolutionised the treatment of bacterial diseases. In the 1940s it was envisaged that toothpaste and even lipstick would be impregnated with antibiotics, protecting us from infection. Antibiotics were thought to have the potential to eradicate most of the major infectious diseases. Why has this proved not to be the case? How is the evolution of microbes causing drug developers a real problem? Antibiotics remain an important weapon in the fight against infection. So how do they work?

It is possible to avoid infection by taking preventive measures. Being vaccinated as a child or when travelling abroad as an adult may be the answer. Could this solution have helped save our victims?

Overview of the biological principles covered in this topic

In this topic you will look at how a dead body can provide evidence for forensic scientists as they try to identify a person, and decide when and how they died. Using your knowledge of DNA and DNA technology gained in the AS course, you will study DNA fingerprinting.

The principles of succession discussed in Topic 5 underpin techniques often employed by forensic entomologists to determine the time of death. Decomposition and recycling by microorganisms discussed in Topic 4 also provides evidence for how long a person has been dead.

You will revisit the structure of prokaryotic cells in the context of pathogens, and look at the symptoms experienced by people infected by TB and HIV. You will gain a detailed understanding of the immune system's role in the body's response to infection. The function of fever in the immune response provides an insight into the use of negative feedback mechanisms for homeostasis.

The principles of evolution are used to consider how pathogens and their hosts (those they infect) are in an evolutionary race. You will examine the evolution of antibiotic resistance in bacteria and see how drug developers are faced with a battle to find new treatments.

6.1 Forensic biology

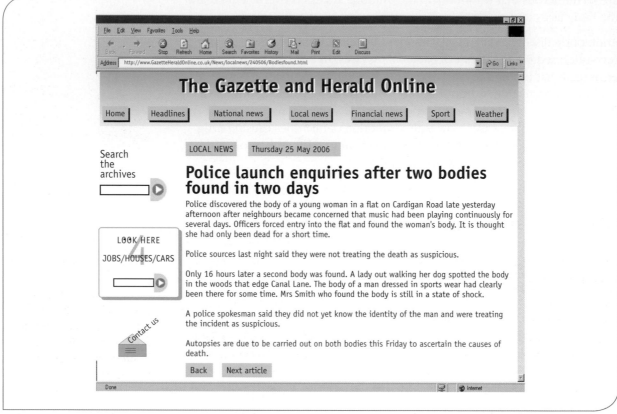

File Edit View Favorites Tools Help

Back Forward Stop Refresh Home Search Favorites History Mail Print Edit Discuss

Address http://www.GazetteHeraldOnline.co.uk/News/localnews/240506/Bodiesfound.html Go Links

The Gazette and Herald Online

Home   Headlines   National news   Local news   Financial news   Sport   Weather

Search the archives

LOOK HERE

JOBS/HOUSES/CARS

Contact us

LOCAL NEWS   Thursday 25 May 2006

Police launch enquiries after two bodies found in two days

Police discovered the body of a young woman in a flat on Cardigan Road late yesterday afternoon after neighbours became concerned that music had been playing continuously for several days. Officers forced entry into the flat and found the woman's body. It is thought she had only been dead for a short time.

Police sources last night said they were not treating the death as suspicious.

Only 16 hours later a second body was found. A lady out walking her dog spotted the body in the woods that edge Canal Lane. The body of a man dressed in sports wear had clearly been there for some time. Mrs Smith who found the body is still in a state of shock.

A police spokesman said they did not yet know the identity of the man and were treating the incident as suspicious.

Autopsies are due to be carried out on both bodies this Friday to ascertain the causes of death.

Back   Next article

Done   Internet

Figure 6.2 Bodies found.

On finding any body (Figure 6.2) the police must first answer three basic questions:

1 Who is the dead person?

2 When did the person die?

3 How did the person die?

Forensic biologists use a variety of techniques to help them answer these questions.

Identifying the body

Q6.1 What do you think the police would first do to identify a body?

The identity of a person is normally fairly easy to establish using the papers that most people carry, such as a diary, bank card, bus pass or receipts. The woman found in Cardigan Road was quickly identified from the photo on her driving licence and from other documents found. She was Nicki Overton, 23 years old, just back from a trip to the US and staying in her friend's flat for a few weeks while the friend visited the Far East. However, when a person is found without any such documents, as was the case with the Canal Lane body, forensic techniques may have to be used to ascertain their identity. Conventional fingerprinting, dental records and, increasingly, DNA fingerprinting are employed.

Fingerprints

Fingerprints are found on all humans, and also on some other animals. They are unique to the individual and remain unchanged over a lifetime. Even identical twins have different fingerprints.

The skin over most of the body is relatively smooth. However, the fingers, palms and soles of the feet have small ridges on the skin surface. Each ridge is a fold in the epidermal layer of the skin, as you can see in Figures 6.3 and 6.4. Ridges vary in width and length; they branch and can also join together, forming distinctive patterns of ridges and furrows.

Sweat and oil secretions leave impressions of our fingerprints on any surface we touch. The oils are secreted from **sebaceous glands**. There are no sebaceous glands on the palms or fingers – they are only found associated with hair follicles on other parts of the body. However, the oils are easily transferred to our hands.

Fresh prints are usually visualised using very fine aluminium powders, although magnets and magnetic powders are sometimes used, Figure 6.5.

Ninhydrin has been used for many years to develop fingerprints on absorbent surfaces. It reacts with the amino acids in sweat to produce a purple-coloured fingerprint impression.

Once a fingerprint has been obtained, an expert will examine the print and compare it with other recorded prints. Modern computerised fingerprint analysis allows the process to be automated, and speeds up searching for a match. The UK police use the National Automated Fingerprint Identification System – a national computerised database holds the 'ten print' records of all convicted criminals.

▲ **Figure 6.3** Fingerprints form early in fetal development, usually by the fifth month.

Weblink

Look in detail at slides through the skin on the University of Illinois College of Medicine interactive histology website.

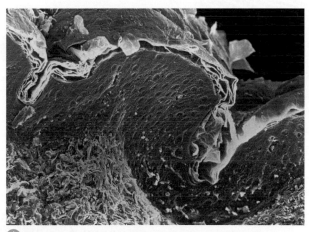

▲ **Figure 6.4** Coloured scanning electron micrograph of ridges in the epidermal tissue of the skin. These ridges form the distinctive pattern of a fingerprint. Magnification ×90.

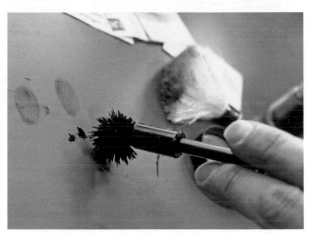

▲ **Figure 6.5** Iron flakes and pigment particles are applied using a magnetic 'brush'. The iron flakes stick to the grease in the fingerprint. The magnet removes excess powder and, because a brush never physically touches the print, a sharper result can be obtained, particularly from difficult surfaces.

Four main types of fingerprint pattern can be identified. This is known as the Henry classification. The four main types – arch, tented arch, whorl and loop – are shown in Figure 6.6. Arch patterns are rare and loop is the most common.

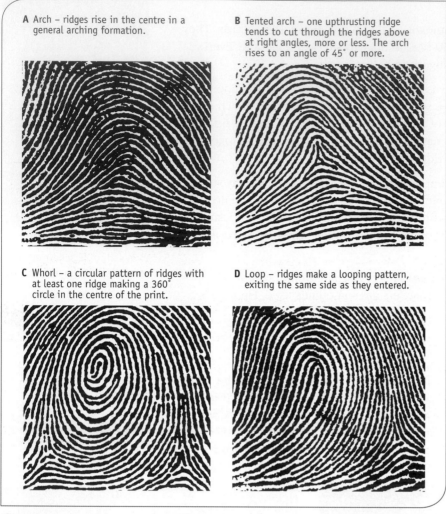

A **Arch** – ridges rise in the centre in a general arching formation.

B **Tented arch** – one upthrusting ridge tends to cut through the ridges above at right angles, more or less. The arch rises to an angle of 45° or more.

C **Whorl** – a circular pattern of ridges with at least one ridge making a 360° circle in the centre of the print.

D **Loop** – ridges make a looping pattern, exiting the same side as they entered.

▲ **Figure 6.6** The four main types of fingerprint pattern.

The matching of these classes alone is not sufficient for identification purposes. A trained examiner looks for minute details in the print that can be used for identification. The positions of branching points, ridge endings and inclusions are important for identification. For a fingerprint match there need to be at least 16 matching points.

Q6.2 Look at Figure 6.6D and identify the points on the print that show:
a a ridge branching **b** a ridge ending **c** inclusion **d** a dot **e** a short ridge.

Q6.3 Fingerprints are not only used for identification of crime suspects. Suggest what other uses could be made of fingerprints.

Q6.4 What evidence could be used to test the hypothesis that there is a genetic contribution to fingerprints?

Q6.5 Fingerprints were taken from the body in Canal Lane. Why were investigators unsuccessful in identifying the body from these prints?

Activity

Activity 6.1 allows you to examine fingerprints in more detail, and find out whether all of a person's fingers have the same pattern. **A6.01S**

Weblink

Listen to the Radio 4 series *Fingerprints* by visiting the BBC website.

Dental records

Dental records may be the best way to identify individuals who have no fingerprints on file, or whose bodies have been damaged in a way that makes other means of identification difficult. Teeth and fillings decay only very slowly and are much more resistant to burning than skin, muscle or bone. Dental records can be as reliable as fingerprints for identifying remains, although records may sometimes include dental work that has not been completed.

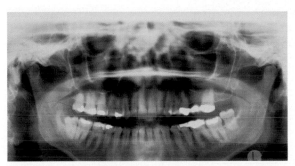

△ **Figure 6.7** Dental X-ray and records. The light marks show fillings, which are highly opaque to X-rays. On the record, CR is a crown and BR is a bridge.

Q6.6 Look at the dental X-ray in Figure 6.7 and match it to the correct person's records. (In reality, a person's actual teeth would be compared with their records.)

A forensic dentist makes an accurate chart of the teeth, including fillings, other dental work and any missing teeth. This can then be compared with dental records of missing persons. Both paper dental records and dental X-rays are useful in the identification process. The forensic dentist may also look at the development of the teeth and roots to help determine the age of the unidentified person in order to narrow down the search.

Dental records were important in the identification of victims of the World Trade Centre attacks in New York on 11 September 2001.

>
> **Activity**
>
> **Activity 6.2** uses dental records and other procedures to identify skeletons of the Russian royal family. **A6.02S**

DNA fingerprinting

DNA fingerprinting (also known as genetic fingerprinting and **DNA profiling**) relies on the fact that apart from identical twins, every person's DNA is unique. As we saw in Topic 3, the human genome contains about 23 000 genes with an average size of 3000 base pairs per gene. This should give about 7×10^7 base pairs in total. However, there are over three billion base pairs (3×10^9) in the genome, so it is clear that a large amount of the DNA does not code for proteins. The non-coding blocks are called **introns** and are inherited in the same way as any gene.

Mini- and micro-satellites

Within introns, short DNA sequences are repeated many times. The sequences of repeated bases are known as mini- or micro-satellites depending on the number of base pairs repeated. They are also known as short tandem repeats.

- **Mini-satellites** usually contain 20–50 base pairs. The mini-satellite can be repeated from 50 to several hundred times.
- **Micro-satellites** usually contain 2–4 base pairs. The micro-satellite can be repeated between 5 and 15 times.

The same satellites occur at the same place (locus) on both chromosomes of a homologous pair. However, the number of times they are repeated on each of the homologous chromosomes can be different, as shown in Figure 6.8. The number of satellite repeats at a locus also varies between individuals.

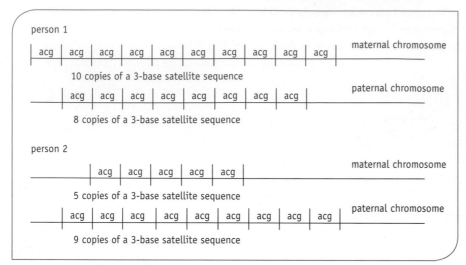

Figure 6.8 Introns at the same locus on each chromosome of a homologous pair may contain different numbers of satellites.

Q6.7 Look at the DNA sequence shown below and identify the micro-satellite sequence. Count how many times it is repeated.

aaccatccttagtaagtaagtaagtaagtaagtattgcccat

Table 6.1 shows how the number of repeats might differ at five satellite loci for two individuals. The large amount of variation in the number of repeats that can occur at each locus, combined with the large number of introns each person possesses, mean that it is highly unlikely that two individuals will have the same combination of satellite repeats.

How is a DNA fingerprint made?

Traditionally, in DNA fingerprinting, a DNA sample is treated with a **restriction enzyme**. This cuts the molecule into fragments at specific base sequences at sites on either side of the satellite repeated sequence. The repeated sequences themselves remain intact, for instance as shown in Figure 6.9.

▽ **Table 6.1** Number of repeats of satellites at five loci for two individuals. A single number at a locus means that the person has the same number of repeats on both chromosomes.

Locus	Person A	Person B
1	15, 14	12, 13
2	5, 9	11
3	10, 7	10, 8
4	7	5, 13
5	8, 10	7, 6

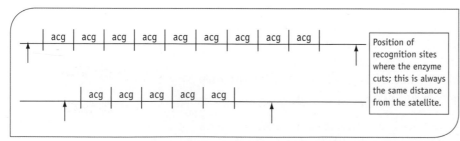

Position of recognition sites where the enzyme cuts; this is always the same distance from the satellite.

◀ **Figure 6.9** Restriction enzymes recognise specific DNA sequences on either side of the satellite repeated sequence.

The fragments are separated by size using **gel electrophoresis**. The fragments' positions on the membrane depend on their size, which in turn depends on the number of repeats in the sequence. The smaller fragments with fewer repeats of the satellite pass through the gel more quickly, so travel further towards the positive electrode. The fragments are then transferred to nylon membrane using **Southern blotting**. This is described in Topic 2 and summarised in Figure 6.10.

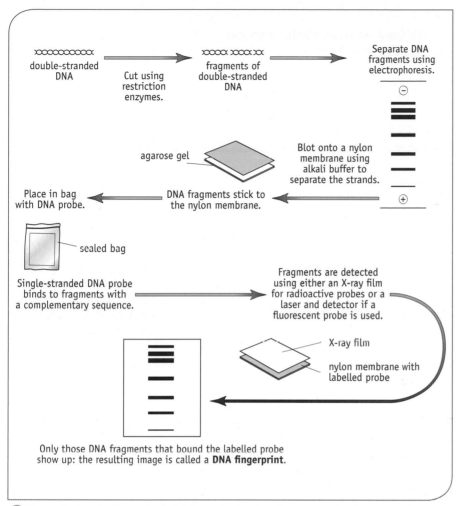

double-stranded
DNA → Cut using restriction enzymes. → fragments of double-stranded DNA → Separate DNA fragments using electrophoresis.

⊖

Blot onto a nylon membrane using alkali buffer to separate the strands.

⊕

agarose gel

Place in bag with DNA probe. ← DNA fragments stick to the nylon membrane.

sealed bag

Single-stranded DNA probe binds to fragments with a complementary sequence. → Fragments are detected using either an X-ray film for radioactive probes or a laser and detector if a fluorescent probe is used.

X-ray film

nylon membrane with labelled probe

Only those DNA fragments that bound the labelled probe show up: the resulting image is called a **DNA fingerprint**.

▲ **Figure 6.10** DNA fingerprints of this type look rather like supermarket bar codes. The Forensic Science Service now uses fluorescent probes to produce a graph-style profile (Figure 6.13).

The membrane is washed with a labelled DNA probe that binds to the repeated sequence producing visible bands. A single band occurs where the maternal and paternal chromosomes have the same number of satellite repeats at the locus. If the two chromosomes each have a different number of repeats there will be two bands.

If this technique is completed using several probes for different repeated sequences, a unique banding pattern for each person is created. The pattern produced is a banded DNA fingerprint such as that shown in Figure 6.10.

Q6.8 Explain what determines the distance travelled by the DNA fragments.

Q6.9 When DNA from person 1 in Figure 6.8 was analysed for the acg repeated sequence, the banding pattern shown in Figure 6.11 was produced. Copy the pattern and sketch where the bands for person 2 would be located relative to these bands.

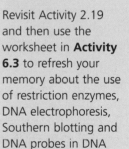

Activity

Revisit Activity 2.19 and then use the worksheet in **Activity 6.3** to refresh your memory about the use of restriction enzymes, DNA electrophoresis, Southern blotting and DNA probes in DNA analysis. **A6.03S**

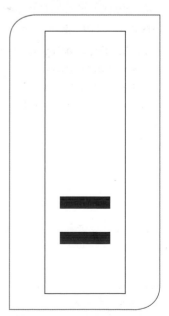

▲ **Figure 6.11** DNA fingerprint.

Polymerase chain reaction

The UK Forensic Science Service uses a **polymerase chain reaction** (PCR) technique and automated gel electrophoresis for DNA fingerprinting, now called DNA profiling.

The PCR reaction rapidly multiplies DNA, allowing forensic scientists to use tiny deposits of hair, skin or body fluid for identification purposes. The process uses DNA primers, short DNA sequences complementary to the DNA next to the satellite that are marked with fluorescent tags. The sample is placed in a reaction tube with **DNA polymerase**, **DNA primers** and nucleotides. Once in the PCR machine the tube is heated and cooled as shown in Figure 6.12. The DNA replicates to produce huge numbers of the DNA fragments.

The Forensic Science Service generally analyses DNA samples for the presence of ten micro-satellites, each four bases in length, with an additional primer pair for determining gender.

Separating the fragments

The fragments of DNA produced by PCR are separated using gel electrophoresis. The system used by the Forensic Science Service is automated. The position of the DNA fragments is revealed as a pattern of fluorescent bands due to the fluorescent tags on the DNA primers attached to the satellites. These bands are detected using a camera or laser scanner.

The results of the electrophoresis are produced as a graph similar to the one in Figure 6.13. The graph is interpreted to give a series of numbers that correspond to the number of repeats in each fragment. This is effectively the DNA fingerprint, now usually called a DNA profile. Table 6.2 shows a digital DNA fingerprint. There are two figures for some of the satellites because the number of repeated sequences is different on each chromosome.

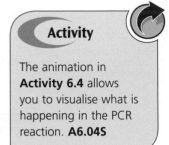

Activity

The animation in **Activity 6.4** allows you to visualise what is happening in the PCR reaction. **A6.04S**

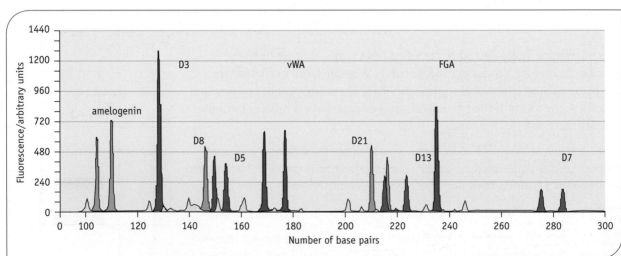

Figure 6.13 A DNA profile. Fluorescence peaks can be converted to a digital readout. *Source: University of Central Lancashire.*

Table 6.2 A DNA profile for six satellites would produce something like this. Amg indicates male or female. A single number means that both chromosomes had the same number of repeats for that sequence.

Amg	D8	FGA	VWA	D13	D18
Y	10	20, 23	16	9, 14	15, 18

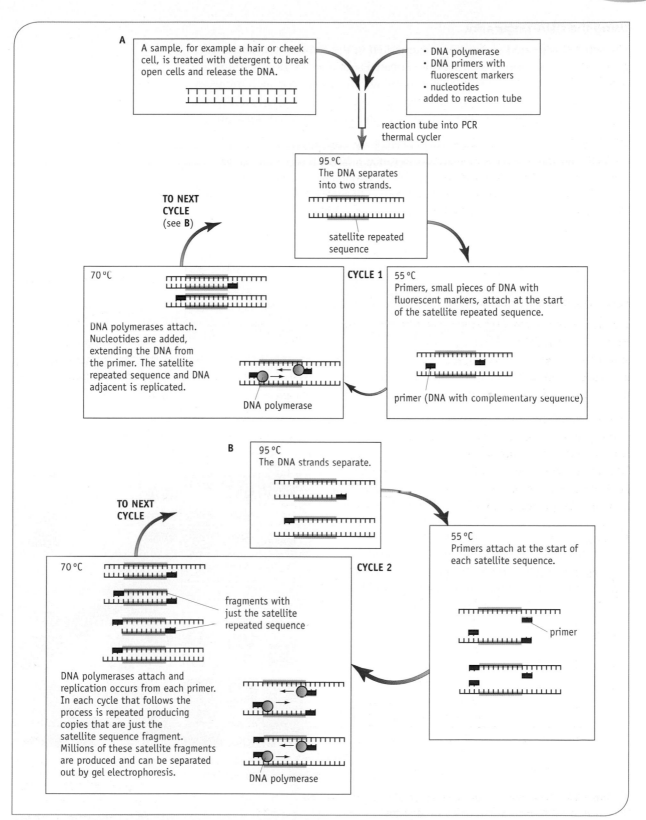

A A sample, for example a hair or cheek cell, is treated with detergent to break open cells and release the DNA.

- DNA polymerase
- DNA primers with fluorescent markers
- nucleotides added to reaction tube

reaction tube into PCR thermal cycler

95 °C
The DNA separates into two strands.

satellite repeated sequence

TO NEXT CYCLE (see **B**)

CYCLE 1

70 °C

DNA polymerases attach. Nucleotides are added, extending the DNA from the primer. The satellite repeated sequence and DNA adjacent is replicated.

DNA polymerase

55 °C
Primers, small pieces of DNA with fluorescent markers, attach at the start of the satellite repeated sequence.

primer (DNA with complementary sequence)

B 95 °C
The DNA strands separate.

TO NEXT CYCLE

CYCLE 2

70 °C

fragments with just the satellite repeated sequence

DNA polymerases attach and replication occurs from each primer. In each cycle that follows the process is repeated producing copies that are just the satellite sequence fragment. Millions of these satellite fragments are produced and can be separated out by gel electrophoresis.

DNA polymerase

55 °C
Primers attach at the start of each satellite sequence.

primer

Figure 6.12 The polymerase chain reaction is used to multiply DNA samples to produce sufficient amounts for separation by gel electrophoresis. The sample goes through many cycles of temperature change, and **A** and **B** show what happens in the first two cycles. The exact temperatures used will depend on which species the DNA comes from. The whole cycle is repeated 25–30 times, taking about 3 hours and producing about 30 million copies of the DNA.

Using the DNA fingerprint

The satellite repeated sequences are inherited in the same way as alleles of a gene, with an offspring receiving one repeated sequence randomly from each parent. This means that genetic fingerprinting can be used for such things as identification purposes, settling paternity disputes, identifying stolen animals, and looking at variation and evolutionary relationships between organisms.

Q6.10 **a** Poppy is a racing pigeon and her chick has been stolen. Her owner, a successful breeder of racing pigeons, suspects that a fellow breeder has stolen the chick and is trying to pass it off as one of his bird Patsy's own brood. Genetic fingerprinting for three satellites produced the results outlined in Table 6.3. Is the chick the offspring of Poppy or of Patsy? Give a reason for your answer.

🔻 **Table 6.3** Digital genetic fingerprints for three pigeons.

| Satellite | Number of repeats for each satellite | | |
	Chick	Poppy	Patsy
A	6, 12	5, 12	10, 6
B	7, 4	4	8, 5
C	9, 5	9, 13	6, 11

b Look at Figure 6.14. The bands in these DNA fingerprints are marked M for mother, F for father and C for child. Decide if the father is the biological parent of both children, giving reasons for your answer.

Are DNA fingerprints infallible?

Genetic fingerprinting has been widely used in legal proceedings to match samples from a crime scene to samples from a suspect. It is generally thought to produce a result that is close to unique to the individual, and a near certain indication of guilt in criminal trials. A complete DNA profile would be unique (even for an identical twin, due to the accumulation of mutations during a person's life) but because the DNA fingerprint test analyses only a few repeated sequences it is less likely to be unique. This is a particular problem if individuals being tested are closely related.

DNA taken from the body in Canal Lane was compared with hair DNA samples collected from the homes of two people who had been reported missing in the local area. The dead man was discovered to be George Watson, a 55-year-old computer software analyst.

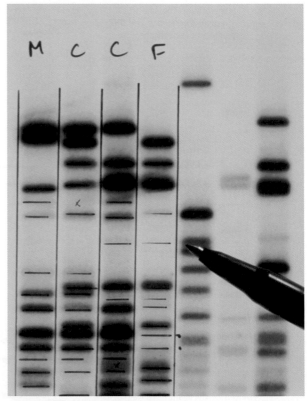

🔺 **Figure 6.14** DNA profiling is frequently used for settling disputes about paternity.

Q6.11 Why would a DNA profile presented as evidence in court include ten or more satellites?

Determining time of death

As soon as a person dies, a series of physical and chemical changes starts to take place in the body. These changes occur in a known order, and can be used to estimate the time of death. The temperature of the body, the degree of **rigor mortis** and the state of decomposition can be used to estimate the time of death. In addition, any entomological evidence provides further clues to help identify the time of death.

Body temperature

Human core body temperature is normally in the range of 36.2–37.6 °C, but as soon as a person dies their body starts to cool due to the absence of heat-producing chemical reactions. The temperature of a body can be useful for estimating time of death within the first 24 hours of death.

Q6.12 The measurement of body temperature to estimate time of death is useful in cool and temperate climates. Why might it be less useful in the tropics?

Core body temperature is measured via the rectum or through an abdominal stab. A long thermometer is needed; an ordinary clinical thermometer is too short and has too small a temperature range. An electronic temperature probe can also be used. The environmental conditions must also be noted, as these will affect how the body has cooled. The cooling of a body follows a **sigmoid curve**, as shown in Figure 6.15. The initial temperature plateau normally lasts between 30 and 60 minutes. The graph assumes that the person's temperature was normal, 37 °C, at the time of death. This may not always be the case. If a person has a fever or is suffering from hypothermia, their body temperature at the point of death will be elevated or depressed.

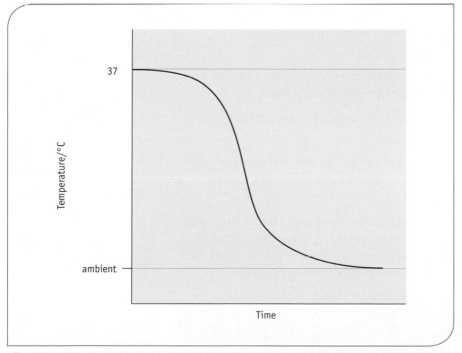

▲ **Figure 6.15** The temperature of a body will follow this sort of sigmoid curve as it cools after death. The initial plateau can last for 30 minutes to several hours depending on conditions.

Over the near linear part of the cooling curve, the temperature decline per hour can be used to give an estimate of the time of death. Many factors will affect post-mortem cooling, including:

- body size
- body position
- clothing
- air movement
- humidity.

If the body is immersed in water it will cool much more rapidly, as water is a better conductor of heat than air. These factors need to be taken into account when estimating time of death from temperature measurements.

Q6.13 Which of the factors mentioned above will slow the rate of cooling?

Q6.14 The Canal Lane body was found to have a core temperature of 18°C. Why might this measurement have been of little use to the forensic pathologists investigating the case?

Rigor mortis

After death, muscles usually totally relax and then stiffen; this stiffening is known as rigor mortis (Latin for 'stiffness of death'). Joints become fixed during rigor mortis and their position, whether bent (flexed) or straight (extended) will depend on the body position at the time of death. After a further period of time, rigor mortis passes and muscles are again relaxed. The following sequence of events occurs.

1 After death, muscle cells become starved of oxygen, and oxygen-dependent reactions stop.

2 Respiration in the cells becomes anaerobic and produces lactic acid.

3 The pH of the cells falls, inhibiting enzymes and thus inhibiting anaerobic respiration.

4 The ATP that is needed for muscle contraction is no longer produced, and as a result bonds between the muscle proteins become fixed.

5 The proteins can no longer move over one another to shorten the muscle, fixing the muscle and joints.

There is a progression in the development of rigor mortis, with smaller muscles stiffening before larger ones. The rigor mortis passes in the same order in which it developed as the muscles start to break down.

Most bodies will have complete rigor mortis between 6 and 9 hours after death and a simple rule of thumb exists (Table 6.4). However, if the environmental temperature is high or the person has been physically active before death then the rigor mortis will set in more quickly and last for a shorter period.

▼ **Table 6.4** A rough guide to the onset and loss of rigor mortis in temperate climates.

Temperature of body	Stiffness of body	Approximate time since death
warm	not stiff	no more than 3 hours
warm	stiff	3–8 hours
cold	stiff	8–36 hours
cold	not stiff	more than 36–48 hours

Decomposition

After death, tissues start to break down due to the action of enzymes. **Autolysis** occurs first. In this process the body's own enzymes, from the digestive tract and from lysosomes, break down cells. (See Topic 3 and the interactive cell in Activity 3.1 to remind yourself of the function of lysosomes.)

Bacteria from the gut and gas exchange system rapidly invade the tissues after death, releasing enzymes that result in decomposition. The loss of oxygen in the tissues favours the growth of anaerobic bacteria.

Signs of decomposition

The first sign of decomposition (also known as putrefaction) in humans is a greenish discoloration of the skin of the lower abdomen. This discoloration will spread across the rest of the body, darken to reddish green and then turn a purple-black colour. Gas or liquid blisters may appear on the skin. Due to the action of bacteria, gases including hydrogen sulphide, methane, carbon dioxide, ammonia and hydrogen form in the intestines and tissues, causing the body to become bloated. As the tissues further decompose the gas is released and the body deflates. When the fluid associated with putrefaction drains away, the soft tissues shrink and the decay rate of the dry body is reduced.

There is variation in the time taken for decomposition. On average, the discoloration of the abdominal wall will occur between 36 and 72 hours after death. Gas formation occurs after about a week. The temperature of the body will determine the rate of decomposition. At higher temperatures, around 26–30°C, gas formation occurs within about three days.

Environmental temperatures have a major influence on the rate of decomposition. Warm temperatures speed up decomposition whereas intense heat denatures the enzymes involved in autolysis, delaying the start of decay. Heavy clothing or coverings over the body increase the body temperature and speed up decomposition. The rate of decomposition is highest between 21 and 38°C. Injuries to the body allow entry of bacteria that aid decomposition.

Q6.15 Which of the following conditions will speed up decomposition and which will slow the process? Give a reason for your answer in each case.
a a well heated room
b injuries to the body surface
c intense heat

Forensic entomology

When the dog walker found the body on Canal Lane it was already discoloured over much of the torso, and a gash on the body was infested with maggots. This was a shocking experience for the person discovering the body, but provided valuable clues for the pathologists trying to determine the time of death.

Q6.16 Using the signs of decomposition described above estimate how long ago the Canal Lane body died.

The presence of insects allows a forensic entomologist to make a more precise estimate of how much time has elapsed since death.

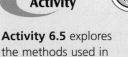

Activity

Activity 6.5 explores the methods used in forensic investigations. **A6.05S**

Forensic entomologists record information about the location and condition of the body. They take samples of any insects found on, near or under the body, noting exactly where and when they were found. The temperature of the air, ground, body and 'maggot mass' are measured, in order that the rate of maggot development can be determined. Some of the maggots will be killed at the time of collection so that their age can be determined later, and some will be taken back to the lab. The live maggots are fed on meat, allowing them to complete their development (Figure 6.16); this is useful for identification of the species.

Q6.17 Explain how an entomologist would identify the species of fly found on a body.

Estimating time since death

There are several ways in which the forensic entomologist can determine the age of maggots and hence estimate the time when the eggs were laid. This provides a minimum time since death:

- For the most common bluebottle species found on bodies, *Calliphora vicina*, the age of the maggot can be read from a graph like the one shown in Figure 6.17. This method can be used only if the temperature conditions of the body have remained fairly constant.

⬆ **Figure 6.16** *Calliphora* sp., a bluebottle fly.

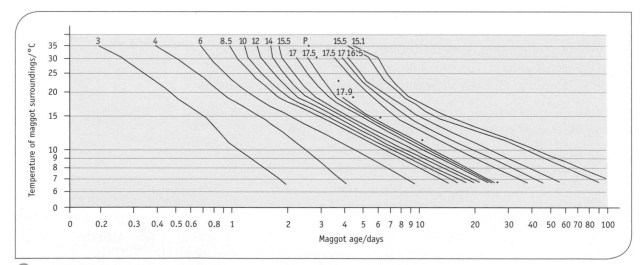

⬆ **Figure 6.17** This graph was produced from a detailed laboratory study of the bluebottle *Calliphora vicina*. Each line is for a maggot of a particular length, the length being given in mm at the top left of the line. Using the temperature of the place where the maggot was found and referring to the appropriate maggot length line, the maggot age can be read from the x-axis. For example, a maggot of 3 mm in length found on a body at 28 °C will be 0.3 of a day or 8 hours old. Lines to the right of the dotted line labelled P are for pupa lengths.

- Identifying the maggot's stage of development with reference to the life cycle of the fly can also give an estimate of age (see Figure 6.18).

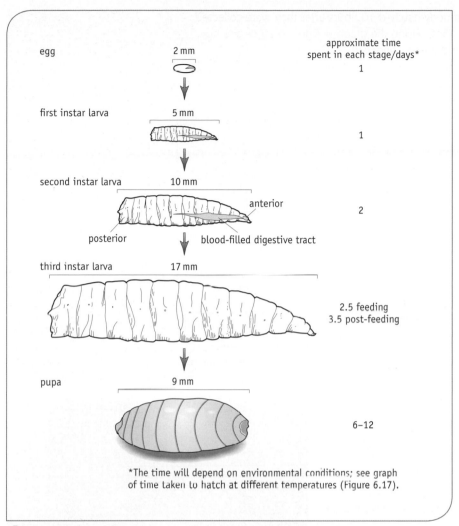

approximate time
spent in each stage/days*

egg   2 mm   1

first instar larva   5 mm   1

second instar larva   10 mm   2

anterior

posterior   blood-filled digestive tract

third instar larva   17 mm   2.5 feeding
3.5 post-feeding

pupa   9 mm   6–12

*The time will depend on environmental conditions; see graph
of time taken to hatch at different temperatures (Figure 6.17).

Figure 6.18 Blowflies (family Calliphoridae) have four life stages: egg, larva (maggot), pupa and adult. There are three larval stages, known as instars. At the end of each instar, the maggot sheds its skin to allow more growth in the next instar. The pupa forms inside a dark hard case.

- If the maggot is not *C. vicina* and its stage of development is unclear, it can be allowed to mature. This gives a date of pupation. If the normal length of time for this species of egg to develop and pupate is subtracted from the date of pupation it should be possible to work out the date on which the eggs were laid. The time of egg-laying may give an underestimate of time of death, because there is no knowing how long it took the flies to find the body. Generally, however, insects will lay eggs on a body within two days of death. In addition, other factors such as toxins in the body can affect maggot development. For example, cocaine would stimulate development.

Q6.18 The bluebottle maggot in Figure 6.19 was collected from the Canal Lane body. Determine its approximate age using the graph in Figure 6.17 (see Question 6.14 for the body temperature).

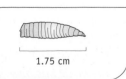

1.75 cm

Figure 6.19 A maggot collected from the Canal Lane body.

Q6.19 The eggs collected on a body hatched 10 hours after they were collected and the body temperature was 15°C. Using the graph in Figure 6.20 decide how long before the body was found the eggs were laid.

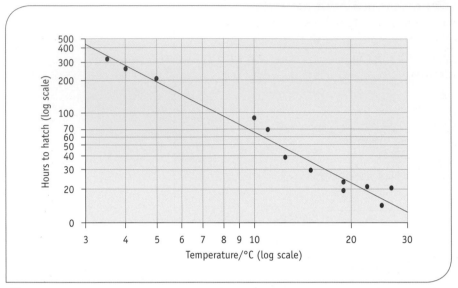

🔺 **Figure 6.20** The time taken for eggs of *Calliphora vicina* to hatch at different temperatures. *Source: The Natural History Museum.*

Q6.20 The body of Kathleen McClung was found in Guildford in June 1969. The stage of development of *Calliphora vicina* maggots on the body could not be clearly identified, so they were allowed to develop on raw beef. Pupae were produced between 4 and 8 July. With reference to the information in Figure 6.18, decide when the bluebottle eggs were laid and estimate the date of death.

Succession on corpses

A corpse attracts many different types of insect. Some of the insects feed off the decaying corpse, and others are attracted to feed off the insects around it. In the 1960s, scientists began to make sense of the range of organisms found on corpses by using the concept of **succession** as developed in ecology. M. Lee Goff, a leading forensic entomologist, explained such succession as:

the idea that as each organism or group of organisms feeds on a body, it changes the body. This change in turn makes the body attractive to another group of organisms, which changes the body for the next group, and so on until the body has been reduced to a skeleton. This is a predictable process, with different groups of organisms occupying the decomposing body at different times.

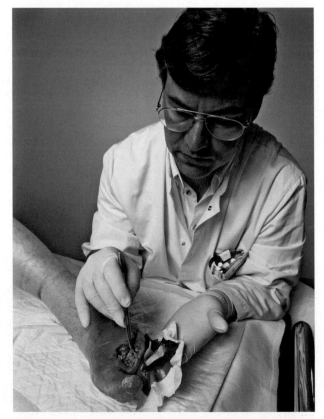

🔺 **Figure 6.21** Maggots are increasingly being used to clean wounds. Maggots of the fly *Lucilia sericata* are bred in sterile conditions. Placed on a wound they eat dead and dying flesh, leaving the healthy tissue. Maggot saliva has antiseptic properties.

Oxford & Cherwell Valley College
Learning Resources

Table 6.5 gives some examples of the species that occur in the succession on a decomposing body. Normally eggs are laid in wounds or at the openings to the body, for example the mouth or nose. The community of species that first occupies the body will be determined by the prevailing conditions. Some species such as *Calliphora vicina* are found in more urban situations, whereas *Calliphora vomitoria* occurs in rural locations, and is rarely found on bodies indoors. Bluebottles lay their eggs in more shaded situations, whereas greenbottles prefer sunny spots.

▼ **Table 6.5** Species succession on a human body.

	Organisms	State of body
First wave	bluebottles: *Calliphora vicina* *Calliphora vomitoria* greenbottle *Lucilia sericata* (sheep blowfly) house fly *Musca domestica* cow face fly *Musca autumnalis*	fresh
Second wave	flesh flies *Sarcophaga* spp.	bloated by gases
Third wave	*Dermestes* beetle larvae *Aglossa* tabby moth maggot	active decomposition (fatty acids have turned to a waxy substance)
Fourth wave	cheese skippers *Piophila* spp. lesser house fly *Fannia canicularis*	active decomposition (fermentation)

Table 6.6 shows how the number of species increases through the stages of the succession. The length of each stage in the succession depends on the condition of the body, which in turn depends on environmental conditions. Note, though, that unlike plant succession, where many of the early species are replaced as conditions change, most of the early insects *remain* on the body.

Q6.21 **a** Explain how temperature might affect the processes of succession on a human body.
b What other factors might affect succession?

▼ **Table 6.6** Total number and percentage of species attracted to the different stages of decay.

Stage of decomposition	Total number of species attracted to each stage of decomposition	Percentage of species attracted to another stage of decomposition				
		Fresh	Bloated	Active decay	Advanced decay	Dry
fresh	17	100	94	94	76	0
bloated	48	33	100	100	90	2
active decay	255	6	19	100	98	13
advanced decay	426	3	10	59	100	38
dry	211	0	1	16	76	100

Insects can also help determine whether a body has been moved. There may be species of insect found on a body that would not naturally occur in that location. For example, insects normally found in woods occurring on a body discovered indoors would suggest that it has been moved some time after the time of death.

Checkpoint

6.1 Produce a bullet-point summary of methods that can be used to determine time of death.

6.2 Cause of death

A post-mortem examination may be performed if there is a sudden or unexpected death, or if the cause of death is unknown. The pathologist will first make an external examination.

An internal examination is then undertaken. An incision is made down the front of the body and organs are taken out for detailed study (Figure 6.22). The state of the internal organs may allow conclusions to be made about the health of the person and any illness suffered. For example, the condition of the heart and arteries may show whether atherosclerosis and a heart attack were responsible for death. Cirrhosis of the liver, characterised by death of liver cells and formation of fibrous tissue, may suggest inadequate diet, excessive alcohol consumption or infection.

Blood and tissue samples may be taken and tested for toxins, infection or tumours. The contents of the stomach may be analysed. This can show what was last eaten and when. This may help in determining when the person was alive and where they were.

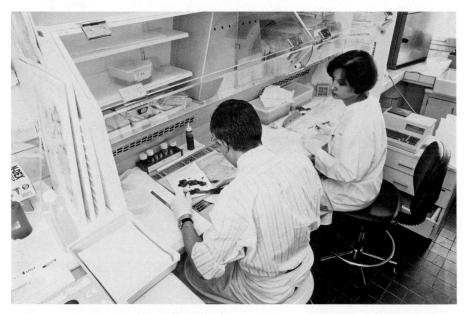

◀ **Figure 6.22** Medical laboratory scientific officers examine lung tissue.

What killed George Watson and Nicki Overton?

The post-mortem on George Watson, the Canal Lane body, showed that he had an aneurysm in the aorta. As we saw in Topic 1, blood builds up behind a section of the artery which has narrowed and become less flexible; the artery bulges as it fills with blood.

Q6.22 **a** What process is likely to have caused the narrowing of the artery?
b What major risk factors could have contributed to the development of the aneurysm?
c What may have happened to the aneurysm, and why would it have caused the man's death?

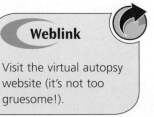

Weblink

Visit the virtual autopsy website (it's not too gruesome!).

Results of Nicki Overton's post-mortem showed she was infected with two pathogenic (disease-causing) microorganisms, *Mycobacterium tuberculosis*, the bacterium responsible for the disease **tuberculosis** (**TB**), and the **human immunodeficiency virus** (**HIV**), the virus that causes **AIDS**. What is the difference between these two microbes? How might she have contracted the infections? Could one or other, or both, have caused her death? To determine whether they were the cause of her death, the extent of her illness and the immune response of the body must be ascertained.

Key biological principle: What is the difference between a virus and a bacterium?

Bacteria

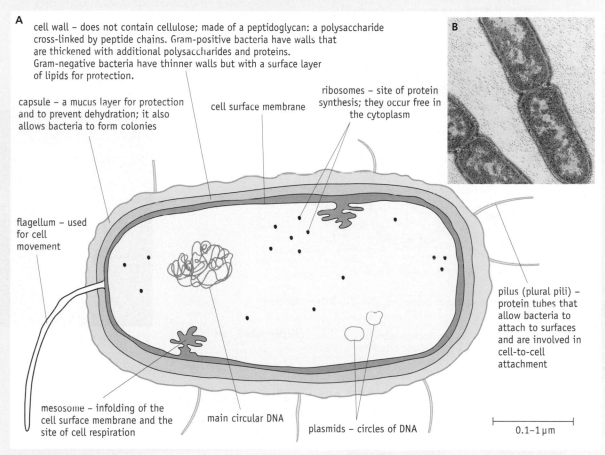

A
cell wall – does not contain cellulose; made of a peptidoglycan: a polysaccharide cross-linked by peptide chains. Gram-positive bacteria have walls that are thickened with additional polysaccharides and proteins. Gram-negative bacteria have thinner walls but with a surface layer of lipids for protection.

capsule – a mucus layer for protection and to prevent dehydration; it also allows bacteria to form colonies

cell surface membrane

ribosomes – site of protein synthesis; they occur free in the cytoplasm

B

flagellum – used for cell movement

pilus (plural pili) – protein tubes that allow bacteria to attach to surfaces and are involved in cell-to-cell attachment

mesosome – infolding of the cell surface membrane and the site of cell respiration

main circular DNA

plasmids – circles of DNA

0.1–1 μm

🔺 **Figure 6.23 A** The basic structure of a bacterium. **B** *Mycobacterium tuberculosis*, first identified by Robert Koch in 1882. Magnification ×5200.

As we saw in Topic 3, bacteria are prokaryotic cells; these are much simpler than eukaryotic cells (Figure 6.23). They do not have a nucleus, lack membrane-bound organelles, and do not produce a spindle during cell division. The average diameter of a bacterial cell is between 0.5 and 5 μm. Bacteria reproduce asexually by binary fission; after replication of the DNA they divide into two identical cells.

Q6.23 Which of the following features would be found in a prokaryote?
a cytoplasm
b a long circular strand of DNA
c plasmids
d cell surface membrane
e mitochondria
f mesosomes (infoldings of the cell surface membrane)

(Continued)

Viruses

Viruses are small organic particles with a much simpler structure than bacteria. They consist of a strand of nucleic acid (RNA or DNA) enclosed within a protein coat (Figure 6.24). Viral DNA can be single or double stranded.

Some viruses also have an outer envelope taken from the **host** cell's surface membrane; the envelope therefore contains lipids and proteins. Viral envelopes also have glycoproteins from the virus itself – these are **antigens**, molecules recognised by the host's immune system as not being its own self. This envelope helps the virus attach to the cell and penetrate the surface membrane. The human immunodeficiency virus is an example of an enveloped virus (Figure 6.48).

Viruses come in a wide variety of sizes, and shapes of different complexity (Figures 6.24 and 6.25).

Viruses lack some of the internal structures required for growth and reproduction. This means they have to enter the cells of the organism they infect (the host) and use the host's metabolic systems to make more viruses, as summarised in Figure 6.26. When viruses hijack the host cell's biochemistry, this disrupts the normal working of the cell. After reproducing inside the host cell, new virus particles may bud from the cell surface or burst out of the cell, splitting it open. This splitting kills the cell and is called **lysis**. It results in the cell contents being released into the surrounding tissues; the many enzymes and other chemicals released can damage neighbouring cells. These processes cause the disease symptoms produced by the virus infection.

A Basic structure

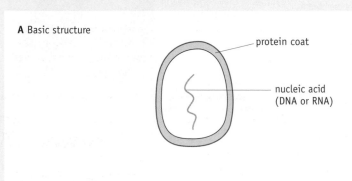

protein coat

nucleic acid (DNA or RNA)

B Viruses come in a wide variety of shapes and sizes (10–300 nm diameter).

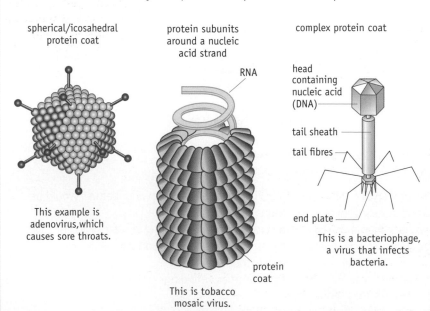

spherical/icosahedral protein coat

This example is adenovirus, which causes sore throats.

protein subunits around a nucleic acid strand

RNA

protein coat

This is tobacco mosaic virus.

complex protein coat

head containing nucleic acid (DNA)

tail sheath

tail fibres

end plate

This is a bacteriophage, a virus that infects bacteria.

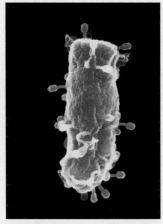

▲ **Figure 6.25** Electron micrograph of bacteriophage viruses attacking an *Escherichia coli* bacterium. Magnification ×17 000.

◀ **Figure 6.24 A** The basic structure of a virus. **B** Some examples of different viruses.

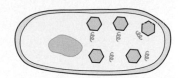

1 Virus attaches to the host cell.

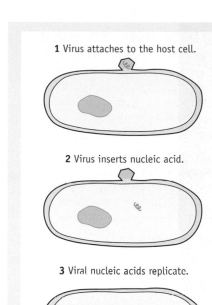

2 Virus inserts nucleic acid.

3 Viral nucleic acids replicate.

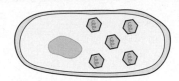

4 Viral protein coats synthesised.

5 New virus particles formed.

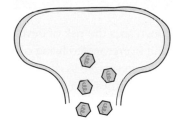

6 Virus particles released due to cell lysis.

Activity

In **Activity 6.6** you can produce an animation brief for a new prokaryote section for the virtual cell. **A6.06S**

Figure 6.26 Viruses use the host cell's protein-synthesis machinery to manufacture new virus particles.

Bacteria and viruses that cause diseases are known as **pathogens**. Common diseases caused by bacteria include *Salmonella* food poisoning, gonorrhoea and cholera. Viruses cause many human diseases including flu, measles, chickenpox and cold sores. Many plant diseases are also the result of viral infection. For example tobacco mosaic virus, as its name suggests, causes mosaic disease in tobacco plants.

Checkpoint

6.2 Draw up a table of comparison to allow you to distinguish between the structure of bacteria and that of viruses.

Q6.24   Which of the following diseases are caused by bacteria and which by viruses?
a common cold
b food poisoning
c cold sores
d cholera
e rubella

Q6.25   Are viruses living or non-living?

Q6.26   Calculate approximately how much smaller the viruses shown in Figure 6.25 are than *Mycobacterium tuberculosis*, the bacterium shown in Figure 6.23B.

How might Nicki have become infected?

Transmission of the TB bacterium

Mycobacterium tuberculosis is carried in the droplets of mucus and saliva released into the air when an infected person talks, coughs or sneezes (Figure 6.27). Others then inhale the droplets. This is known as droplet infection. The droplets can remain suspended for several hours in poorly ventilated areas.

Figure 6.27 TB bacteria are carried in the droplets released by an infected person, for example when they sneeze.

Close contact with an infected person increases the risk of developing the disease, as do poor health, poor diet and overcrowded living conditions. *M. tuberculosis* is a tough bacterium and can survive as dust from dried droplets for several weeks, making the bedclothes and room of a TB patient potentially infectious.

Transmission of HIV

HIV is not a very tough virus and cannot survive outside the body for any significant time. It can be passed on only in body fluids, such as blood, vaginal secretions and semen, but not saliva or urine. Infection can occur if you have unprotected sex with someone who is already infected, or if blood from such a person enters your bloodstream. For infection to occur, the body fluids have to be transferred directly into the body of the next host. This can occur in the following ways:

- Infection can result from sharing needles, whether used illegally for drugs or legally.
- Through unprotected sex. Worldwide, this is the most frequent route of infection. The virus can enter the bloodstream of a partner through breaks in the skin or lesions caused by other infections (usually other **sexually transmitted infections, STIs**). The use of a condom can prevent this transmission. Infection can also occur, though rarely, via oral sex.
- Direct blood-to-blood transfer can occur through cuts and grazes. Police, paramedics and medical staff are particularly at risk from this method of transmission, and precautions are taken to minimise the risk.
- Maternal transmission can occur from mother to unborn child or in breast milk. The risk of the virus being passed to the baby occurs in the last few weeks of pregnancy and mostly around the birth itself, when mingling of infant and maternal blood is likely to happen. Taking anti-HIV drugs during the last three months of pregnancy and giving birth by Caesarean section greatly reduce this risk, from about 20% to 5%. This option is only realistically available in countries with advanced medical care.

Q6.27 Suggest what prevents HIV from infecting a baby during most of the pregnancy.

6.3 How does the body respond to infection?

When someone is infected by a disease-causing organism, several mechanisms come into action in their body to attempt to destroy the invading pathogen. This task is performed by the **immune system**, and is known as the **immune response**.

Non-specific responses help to destroy any invading pathogen, whereas specific immunity is always directed at a specific pathogen. The immune response is summarised in Figure 6.28.

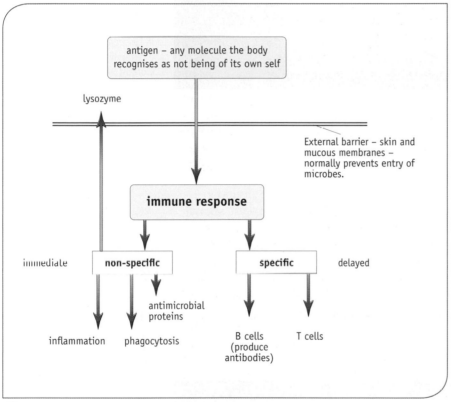

▲ **Figure 6.28** The immune response.

Non-specific responses to infection

Lysozyme

If a speck of dirt lands in your eye, your eye starts to weep. The stream of tears helps to wash out the foreign material and attacks any bacteria that may have got onto the surface of the eye. Tears contain an enzyme called lysozyme (Figure 6.29) that kills bacteria by breaking down their cell walls. The same enzyme is found in saliva and nasal secretions, protecting the body from harmful bacteria which may be present in the air we breathe or the food we eat.

Q6.28 Suggest how the enzyme lysozyme might break down bacterial cell walls. (Think about the structure of bacterial cell walls before answering the question.)

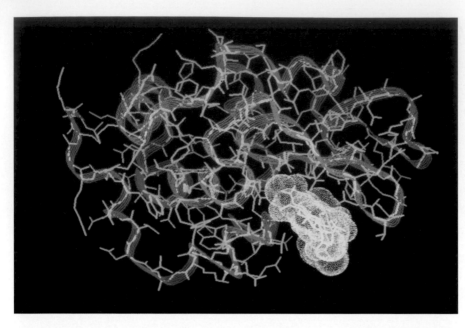

Figure 6.29 The molecular structure of lysozyme was determined in 1965. This was the first time the structure of an enzyme had ever been worked out. Lysozyme breaks down the cell walls of bacteria, and is important in defending the body against infection.

Inflammation

An injury, cut or graze enables microbes and other foreign material to enter the body. A blood clot will rapidly seal the wound but inflammation at the site, known as the **inflammatory response**, is involved in the destruction of the microbes that entered. Damaged white blood cells and mast cells, found in the connective tissue below the skin and around blood vessels, release special chemicals such as **histamine**. These chemicals cause the arterioles in the area to dilate, increasing blood flow in the capillaries at the infected site. Histamines also increase the permeability of the capillaries: cells in the capillary walls separate slightly, so the vessels become leaky. Plasma fluid, white blood cells and antibodies leak from the blood into the tissue causing **oedema** (swelling). The infecting microbes can now be attacked by other intact white cells.

Q6.29 Explain why the finger shown in Figure 6.30 became hot, red and swollen after the man cut his finger.

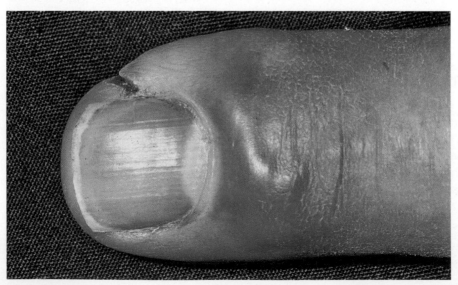

Figure 6.30 This finger became infected after the man cut himself with a bread knife.

People who get hay fever have an **allergic reaction** to certain types of pollen. The nasal tissues respond to the presence of pollen grains by releasing large amounts of histamine. This causes the nasal lining to swell and itch. It also becomes leaky, giving the characteristic runny nose. Taking antihistamine tablets greatly reduces this allergic response.

Antihistamines and histamines have similar shapes. Antihistamines bind to histamine receptors and block the binding of histamines.

Phagocytosis

Phagocytes are white blood cells that engulf bacteria and other foreign matter in the blood and tissues. Phagocytes include both **neutrophils** and **macrophages**. You can see the difference between the different types of blood cell in Figure 6.31.

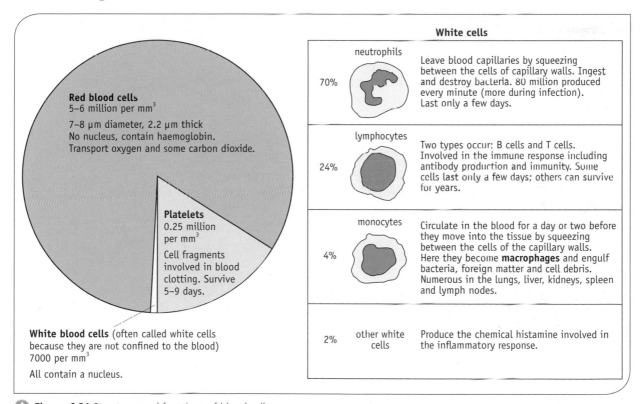

Red blood cells
5–6 million per mm^3

7–8 μm diameter, 2.2 μm thick
No nucleus, contain haemoglobin.
Transport oxygen and some carbon dioxide.

Platelets
0.25 million per mm^3

Cell fragments involved in blood clotting. Survive 5–9 days.

White blood cells (often called white cells because they are not confined to the blood)
7000 per mm^3

All contain a nucleus.

White cells

70%	neutrophils	Leave blood capillaries by squeezing between the cells of capillary walls. Ingest and destroy bacteria. 80 million produced every minute (more during infection). Last only a few days.
24%	lymphocytes	Two types occur: B cells and T cells. Involved in the immune response including antibody production and immunity. Some cells last only a few days; others can survive for years.
4%	monocytes	Circulate in the blood for a day or two before they move into the tissue by squeezing between the cells of the capillary walls. Here they become **macrophages** and engulf bacteria, foreign matter and cell debris. Numerous in the lungs, liver, kidneys, spleen and lymph nodes.
2%	other white cells	Produce the chemical histamine involved in the inflammatory response.

Figure 6.31 Structures and functions of blood cells.

Action at the infected site

Chemicals released by bacteria and the cells damaged at the site of infection attract phagocytic white cells. Neutrophils are the first to arrive; they engulf between 5 and 20 bacteria before they become inactive and die (Figure 6.32). The neutrophils are followed by macrophages. These larger, longer-lived cells each have the potential to destroy as many as 100 bacteria. They will also ingest debris from damaged cells, and foreign matter such as particles

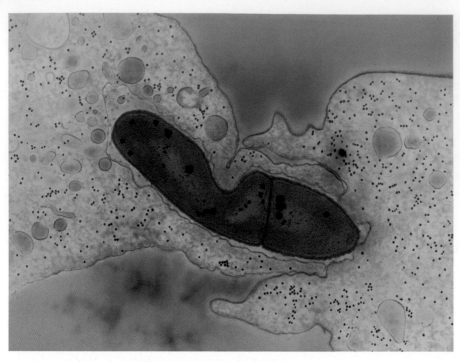

Figure 6.32 Two white blood cells (coloured blue) attempting to engulf a *Bacillus cereus* bacterium. These soil bacteria are a cause of food poisoning. Magnification ×25 000.

of carbon and dust in the lungs. The ingested material is enclosed within a vacuole (Figure 6.33). Lysosomes containing digestive enzymes fuse with the vacuole and the enzymes are released and destroy the bacteria or other foreign material.

The large numbers of phagocytic cells that collect at the site of infection can engulf huge numbers of bacteria. After a few days, the area is full of dead cells, mainly neutrophils, which form a thick fluid called pus. The pus may break through the surface of the skin, but usually it gradually gets broken down and absorbed into the surrounding tissue.

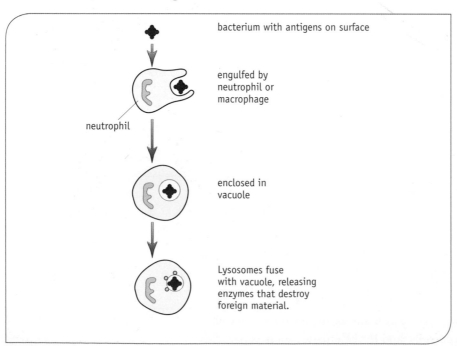

Figure 6.33 At the site of infection neutrophils and macrophages engulf bacteria.

Did you know? Radicals and antioxidants

After engulfing bacteria, phagocytes, especially neutrophils, produce the highly reactive radicals nitric oxide (NO•) and superoxide (O_2•). These kill the engulfed bacteria by attacking their DNA and protein.

As well as helping to defend against infections, radicals can also damage the body's own cells, thereby increasing the risk of diseases such as atherosclerosis. Certain enzymes counteract this effect by destroying any excess radicals. Antioxidants in the diet, for example vitamins C and E, have a similar protective action.

Components in cigarette smoke cause the production of unnaturally high levels of radicals and this is linked to the fact that smokers are much more likely to have heart attacks and strokes.

Action to prevent the spread of infection

In spite of intense phagocytic activity at the infected site, some live bacteria usually get carried away either by the blood or in the lymph. The spread of these bacteria is hindered by the action of macrophages in the **lymph nodes**, spleen and liver. The role of the lymphatic system is outlined in Figure 6.34. Only occasionally does the system fail, leading to widespread infection known as septic shock or 'blood poisoning'.

▼ **Figure 6.34** The location and role of the lymph nodes in the immune response.

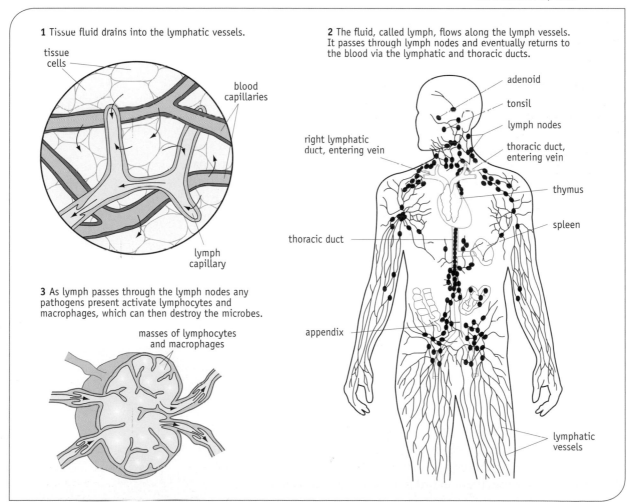

1 Tissue fluid drains into the lymphatic vessels.

tissue cells

blood capillaries

lymph capillary

3 As lymph passes through the lymph nodes any pathogens present activate lymphocytes and macrophages, which can then destroy the microbes.

masses of lymphocytes and macrophages

2 The fluid, called lymph, flows along the lymph vessels. It passes through lymph nodes and eventually returns to the blood via the lymphatic and thoracic ducts.

adenoid

tonsil

lymph nodes

right lymphatic duct, entering vein

thoracic duct, entering vein

thymus

spleen

thoracic duct

appendix

lymphatic vessels

Antimicrobial proteins – interferon

Most of the body's non-specific defences are aimed at invading bacteria. **Interferon** is the exception, as it provides non-specific defence against viruses. Virus-infected cells produce this protein; it diffuses to the surrounding cells where it prevents viruses from multiplying. It acts as a signal molecule inhibiting viral protein synthesis, and in this way limits the formation of new virus particles.

Checkpoint

6.3 Produce a flowchart showing the sequence of events that occurs in the non-specific response at the site of a cut.

Did you know? A miracle drug?

When interferon was first discovered, scientists hoped that it would become a 'miracle drug' for viral diseases and possibly cancer. Chemically, interferon is a small protein existing in a number of species-specific forms. These can be produced artificially using genetically modified bacteria. However, the production is extremely costly and when the drug is injected into patients it lasts only a short time and produces unpleasant side effects. Therefore, although interferon is sometimes used for the treatment of hepatitis, it is almost never used for other viral diseases.

Interferon is used in the treatment of the autoimmune disease multiple sclerosis. Recent research is trying to develop alternative drugs which can boost the natural production of interferon by the body.

Specific immunity

Lymphocytes are white blood cells that help to defend the body against specific diseases (Figure 6.31). They circulate in the blood and lymph and gather in large numbers at the site of any infection.

Reserve supplies of lymphocytes are held in strategically positioned lymphoid tissue (Figure 6.34). If, for example, you have an ear infection, lymphocytes in the lymph nodes of the neck go into action causing 'swollen glands'. If you have an infected finger, it is the lymph nodes in the armpit that become swollen and active. The tonsils and adenoids are well positioned in the throat and nose to help deal with any upper respiratory infections, whereas patches of lymphoid tissue around the stomach and intestine protect against gut infections.

B and T cells

There are two main types of lymphocyte: **B cells** and **T cells**. Both types respond to antigen molecules such as the ones that occur on the surface of bacteria or viruses. Most antigens are protein molecules, and their large size and characteristic molecular shape allow the lymphocytes to identify which ones are 'foreign' (non-self). The response by lymphocytes is called the **specific immune response**.

B lymphocytes

B cells secrete **antibodies** in response to antigens. Antibodies (Figure 6.35) are special protein molecules of a class known as **immunoglobulins**. The antibodies produced bind to antigens and act as a label which enables phagocytes to recognise them in order to destroy them (Figure 6.36).

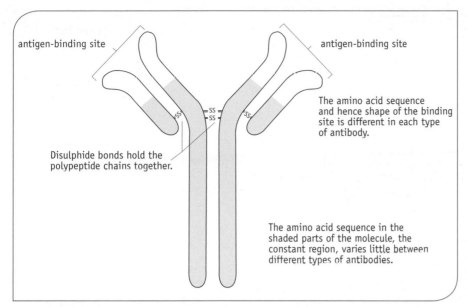

antigen-binding site

antigen-binding site

The amino acid sequence and hence shape of the binding site is different in each type of antibody.

Disulphide bonds hold the polypeptide chains together.

The amino acid sequence in the shaded parts of the molecule, the constant region, varies little between different types of antibodies.

Figure 6.35 A simplified diagram of an antibody. The antigen with a complementary shape can bind to the antibody's antigen-binding site.

Each B cell produces only one type of antibody that can bind to only one specific antigen. A microbe will usually have several antigens on its surface. Each will bind and activate different B cells.

During early human embryo development, 100 million different B cells are produced in the bone marrow. Each of these divides rapidly to produce a clone of cells, providing the baby with an immune system which can respond to a tremendous variety of antigens that might invade its body after birth. B cells have receptors on their surface, and these receptors include transmembrane versions of the antibody molecules they produce.

T lymphocytes

T lymphocytes, like B lymphocytes, are produced in the bone marrow but, unlike B lymphocytes, they mature in the **thymus gland** – hence T for thymus (Figure 6.37). T cells have one specific type of antigen receptor on their surface. This will bind only with the appropriate antigen that has the complementary shape.

There are two types of T cell, T helper cells and T killer cells.

- **T helper cells** – when activated these stimulate the B cells to divide and form cells capable of producing antibodies. They also enhance the activity of phagocytes.

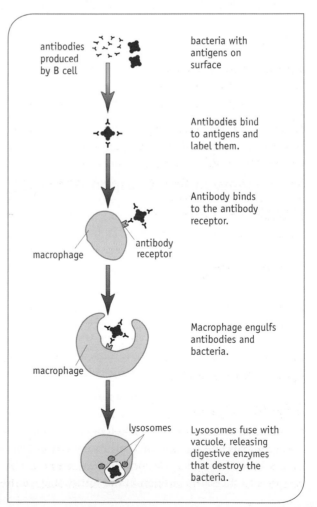

antibodies produced by B cell

bacteria with antigens on surface

Antibodies bind to antigens and label them.

Antibody binds to the antibody receptor.

macrophage

antibody receptor

Macrophage engulfs antibodies and bacteria.

macrophage

lysosomes

Lysosomes fuse with vacuole, releasing digestive enzymes that destroy the bacteria.

Figure 6.36 Antibodies attach to antigens, allowing phagocytes to identify them.

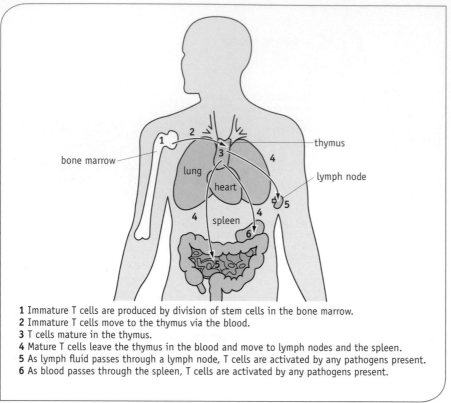

1 Immature T cells are produced by division of stem cells in the bone marrow.
2 Immature T cells move to the thymus via the blood.
3 T cells mature in the thymus.
4 Mature T cells leave the thymus in the blood and move to lymph nodes and the spleen.
5 As lymph fluid passes through a lymph node, T cells are activated by any pathogens present.
6 As blood passes through the spleen, T cells are activated by any pathogens present.

🔺 **Figure 6.37** T cells move from the bone marrow to the thymus, where they mature before passing to lymph tissue via the blood.

- **T killer cells** – these destroy pathogen-infected cells. They also attack any other foreign cells that may enter the body. This causes problems for patients who have undergone transplant surgery since the T killer cells attack the transplant unless it has come from another part of the patient's body.

The primary immune response

Activation of T cells

When a piece of biological material has been engulfed by a macrophage, some of its protein fragments (peptides) are displayed on the surface of the cell (Figure 6.38A). Cell membranes are fluid, and proteins produced by the cell are continually added to and removed from the 'fluid mosaic' of the cell surface membrane. The peptides known as antigens are carried to the plasma membrane by proteins called **major histocompatibility complexes (MHCs)**. When presented in this way the antigens act as a signal to alert the immune system that there are foreign antigens in the body. This cell is now known as an **antigen-presenting cell (APC)**, as shown in Figure 6.38A.

A T helper cell with a complementary shaped receptor, called a **CD4 receptor**, on its surface binds to the antigen fragment displayed by the antigen-presenting cell. This T helper cell divides to produce a clone of active T helper cells and a clone of **T memory cells**. This second clone of cells is longer lived, remaining for months or years in the body. So if an individual is exposed to the same antigen in the future they can respond more quickly.

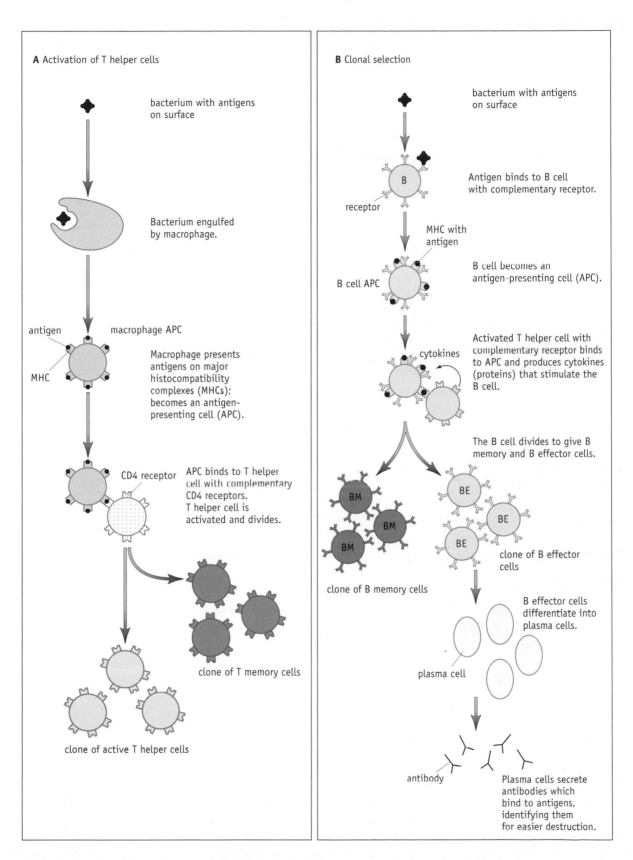

Figure 6.38 A Activation of T cells. **B** Cell clonal selection.

Cloning of B cells

When antigens are detected, B cells with complementary receptors bind to them and become antigen-presenting cells (Figure 6.38B) (B cells engulf and present antigens on MHCs in the same way as macrophages). Active cloned T helper cells that are presenting the same antigen bind to the antigen-presenting B cells. Once attached the T cells release chemicals called **cytokines**, which stimulate division and differentiation of the B cells.

The B cells divide to produce two clones of cells:

- **B effector cells** – these differentiate to produce **plasma cells** that release antibodies into the blood and lymph, but these cells are relatively short lived, lasting only a few days.
- **B memory cells** – these cells live longer, remaining for months or years in the body. So if an individual is exposed to the same antigen in the future they can respond more quickly.

This process is called clonal selection. The first time the B cells are selected by an antigen, the production of sufficient antibody-producing cells takes about 10–17 days; this is the **primary immune response**.

Key biological principle: Complementary protein shapes

You will be familiar with lock-and-key mechanisms for enzyme specificity, but this specificity of binding occurs in many other situations. In Topic 3 we saw that the action of signal proteins within cells depends on the binding of signal molecules with receptors that have complementary shapes. In addition, the formation of the transcription initiation complex relies on the binding together of protein transcription factors.

The immune response relies on the recognition of specific protein antigens by T helper and B cells using receptors with the complementary shape. The specific shape of the binding sites found on antibodies is crucial to their function.

The role of T killer cells

If an intracellular bacterium or virus infects a body cell, a fragment of the antigen is presented on the cell surface membrane via MHCs in the same way as occurs in macrophages. T killer cells with complementary receptors bind to the presented antigen (Figure 6.39). They divide to form a clone of active T killer cells; this division is stimulated by the cytokines from T helper cells. Without cytokines, there would not be enough T killer cells produced to fight a viral infection. The T killer cells destroy cells by releasing enzymes that create pores in the infected cell membrane. Ions and water flow into the infected cell, and it swells and bursts (lysis). The pathogens within the cell are released and can be labelled by antibodies from B cells as targets for destruction by macrophages.

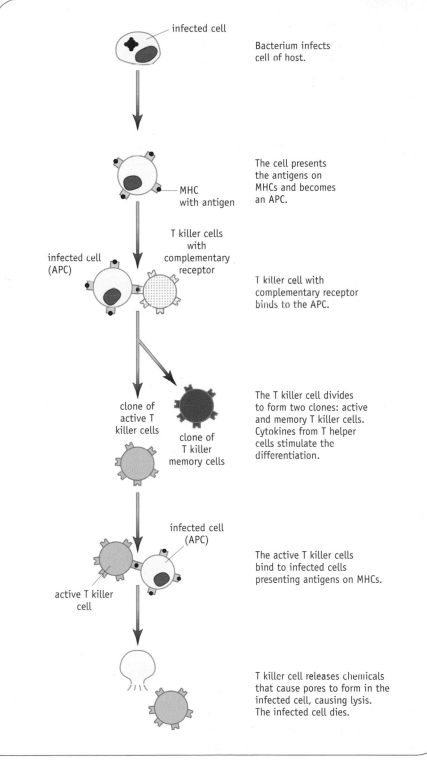

infected cell

Bacterium infects cell of host.

MHC with antigen

The cell presents the antigens on MHCs and becomes an APC.

T killer cells with complementary receptor

infected cell (APC)

T killer cell with complementary receptor binds to the APC.

clone of active T killer cells

clone of T killer memory cells

The T killer cell divides to form two clones: active and memory T killer cells. Cytokines from T helper cells stimulate the differentiation.

infected cell (APC)

active T killer cell

The active T killer cells bind to infected cells presenting antigens on MHCs.

T killer cell releases chemicals that cause pores to form in the infected cell, causing lysis. The infected cell dies.

🔺 **Figure 6.39** The role of T killer cells.

Activities

View the interactive tutorial in **Activity 6.7** to see phagocytosis occurring. **A6.07S**

In **Activity 6.8** an interactive tutorial lets you test your overall understanding of the body's specific immune response to infection and provides you with a summary sheet. **A6.08S**

The secondary immune response

If infected again by the same bacterium or virus, the immune system can respond much faster. The **secondary immune response** involves memory cells and takes only about 2–7 days. The B memory cells produced in the primary response can immediately differentiate to produce plasma cells and release antibodies. There is greater production of antibodies, and the response lasts longer as the second peak in the graph in Figure 6.40 shows. The invading viruses or bacteria are often destroyed so rapidly that the person is unaware of any symptoms. The person is said to be **immune** to the disease.

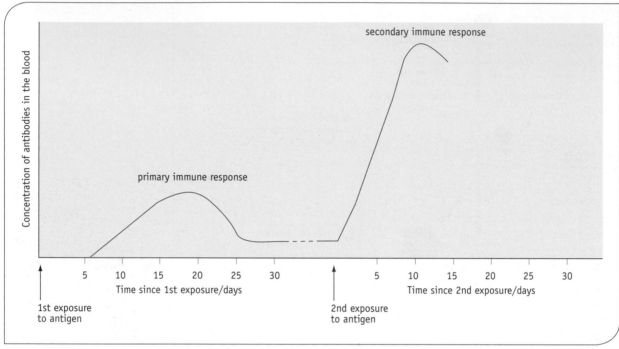

Figure 6.40 Changes in the concentration of antibodies after infection with an antigen.

Q6.30 Which of the B cells will be selected by the antigens in Figure 6.41?

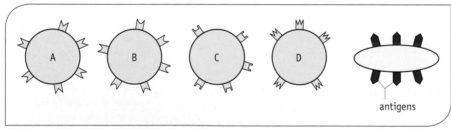

Figure 6.41 Which B lymphocyte will be selected?

Q6.31 The graph in Figure 6.40 shows the antibody concentration in the blood after the first and then second exposure to a foreign antigen. Use the graph to suggest three reasons why people often have no symptoms the second time they encounter a disease.

Checkpoint

6.4 Sketch a diagram or use downloaded figures from the mediabank book artwork to show how Figures 6.36, 6.38A, 6.38B and 6.39 are interconnected.

Avoiding attack by our own immune system

Some of the membrane proteins on the surface of our cells act as 'bar codes'. These proteins, called major histocompatibility complexes (MHCs), mark the body cell as 'self'. They allow us to distinguish between our own cells and those of 'foreign invaders'. There are hundreds of alleles for MHCs, so the combination of MHC proteins on the surface of our cells is unique to each individual and is not found on the cells of anyone else.

As B and T cells are maturing in the bone marrow and thymus, any lymphocytes for 'self' MHCs are destroyed by **apoptosis**, programmed cell death (see Topic 3). Only lymphocytes with receptors for foreign, 'non-self', antigens remain.

Just occasionally the body attacks itself. Particular cells may alter in some way so that they appear 'foreign' and are destroyed by the immune system. An example of this is the auto-destruction of insulin-secreting cells in the pancreas, leading to insulin-dependent diabetes. Other autoimmune diseases are rheumatoid arthritis and multiple sclerosis.

Did you know? A genetically modified immune system

In 2001 gene therapy transformed the life of baby Rhys Evans. The little boy from Cardiff was born with a faulty immune response. He was constantly ill and spent much of his time living in a sterile 'bubble' in London's Great Ormond Street Hospital (Figure 6.42). Rhys had SCID (severe combined immunodeficiency), a rare sex-linked condition caused by a single mutated gene on the X chromosome. Without treatment he would have died within months.

Rhys's defective gene prevented him from producing T killer cells and he was unable to fight off infections. Even chickenpox or cold sores were life threatening.

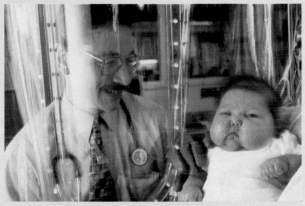

▲ **Figure 6.42** Children with SCID usually have to live their whole lives within a sterile bubble to protect them from infection.

At first Rhys's doctors hoped to give him a bone marrow transplant which would have allowed healthy T helper cells to develop from the donated stem cells. However, they were unable to find a donor whose MHCs were sufficiently similar to prevent rejection of the donated cells.

The only alternative treatment was gene therapy (see Topic 2). Doctors removed bone marrow from Rhys and inserted working versions of his defective gene into the stem cells, using a gibbon virus as the vector. The GM stem cells were then re-implanted into Rhys, where they multiplied and started to generate healthy immune cells. By the time he was a toddler Rhys was out of his 'bubble' and playing with other children. His new GM stem cells should last a lifetime, producing healthy lymphocytes to protect him against infections. In the words of Dr Adrian Thrasher, who led the medical team at Great Ormond Street Hospital: 'We're very excited by this – he was incredibly sick, with a nasty pneumonia, a life-threatening infection. After his gene therapy, he was running around at home – he's a normal little boy now.'

By 2005, Rhys was continuing to do very well and was attending a normal school.

6.4 The body's response to TB

Nicki Overton was infected with TB and HIV. What symptoms might she have had? How might her immune system have responded to each of the pathogens? Could the diseases have been the cause of her death?

What is tuberculosis?

Tuberculosis (TB) is a contagious disease caused by the bacterium *Mycobacterium tuberculosis*. It is an ancient disease: evidence of infection has been found in the bones of 3000-year-old Egyptian mummies. In Western Europe and North America, TB became truly rampant during the nineteenth and early twentieth centuries. The poverty and cramped living conditions of city slums promoted rapid spread of infection, resulting in very large numbers of TB cases. The disease was so active in the population that it was known as the 'white plague', or 'consumption', because it 'consumed' the body of the patient.

The composer Chopin, poet John Keats and the Brontë sisters were among many other famous figures affected by the disease in the nineteenth century. In 1882 Robert Koch said: 'If the number of victims which a disease claims is the measure of its significance, then all diseases ... must rank far behind tuberculosis'.

It is estimated that nearly two billion people – one-third of the world population – is infected. There are estimated to be 8.8 million new cases each year and over two million people will die from the disease every year. Table 6.7 shows the incidence and mortality figures for 2003. However, most infected people are perfectly healthy. Respiratory or pulmonary TB is the most common form. It affects the lungs and is highly contagious.

▼ **Table 6.7** World Health Organisation estimates of TB incidence and mortality in 2003.

WHO region	Number of cases	Deaths from TB	Per 100 000 population
Africa	2 372 000	538 000	78
The Americas	370 000	54 000	6
Eastern Mediterranean	634 000	144 000	28
Europe	439 000	67 000	8
South-east Asia	3 062 000	617 000	39
Western Pacific	1 933 000	327 000	19
Global	8 810 000	1 747 000	28

Improved housing and living conditions, coupled with the development of **antibiotics**, saw a decline in the number of TB cases in the UK and other industrialised nations (Table 6.8) during the twentieth century. However, there has recently been a resurgence of the disease, and although it is still relatively rare in the UK there are approximately 7000 new cases each year with about 500 deaths, mostly in major cities.

Table 6.8 Number of TB cases in England and Wales notified to the Communicable Disease Surveillance Centre. *Source: Office for National Statistics and Health Protection Agency.*

Year	Respiratory	Non-respiratory	Total	Population
1915	68309	22283	90592	35284000
1920	57844	15488	73332	37247000
1925	58545	19228	77773	38935000
1930	50583	16818	67401	39801000
1935	39635	12435	52070	40645000
1940	36151	10421	46572	39889000
1945	42165	9944	52109	37916000
1950	42435	6923	49358	43830000
1955	33580	4554	38134	44441000
1960	20799	2806	23605	45775000
1965	13552	2551	16103	47671200
1970	9475	2426	11901	48891300
1975	8208	2610	10818	49469800
1980	6670	2472	9142	49603000
1985	4660	1197	5857	49990500
1990	3942	1262	5204	50869500
1995	4123	1483	5608	51820200
2000	4825	1747	6572	52943284
2001	4853	1861	6714	52041916
2002	4802	1951	6753	52480500
2003	4585	1933	6518	52793700
2004	4555	2108	6723	53046000

Symptoms of the disease

Only 30% of people closely exposed to TB will become infected, and only 5–10% of those infected will develop the symptoms of the disease. Infection may occur when *M. tuberculosis* bacteria are inhaled and lodge in the lungs. Here they start to multiply. There are two phases to the disease: primary infection (the first phase) and active tuberculosis (the second phase).

Primary infection with TB

The immune system responds

The first phase (primary infection), which can last for several months, may have no symptoms.

The immune system of a person infected with TB responds to and deals with the infection. The *M. tuberculosis* evokes an inflammatory response from the host's immune system. In an individual with a healthy immune system, macrophages engulf the bacteria. A mass of tissue known as a granuloma forms, produced in response to infection. In TB these tissue masses are anaerobic and have dead bacteria and macrophages in the middle (Figure 6.43). They are called tubercules and give the disease its name. After 3–8 weeks, the infection is controlled and the infected region of the lung heals. Most primary infection happens during childhood and over 90% of infections heal without ever being noticed.

Q6.32 *M. tuberculosis* bacteria are obligate aerobes (they need oxygen to survive). Why do they die within tubercules?

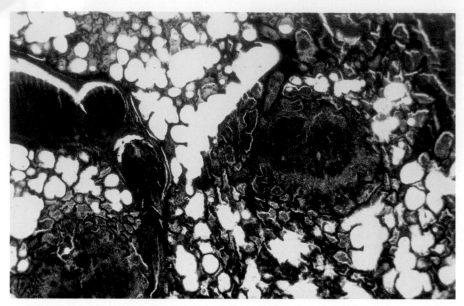

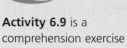
Figure 6.43 Section from a lung of a TB patient. The blue and green areas in the centre and bottom left are tubercules with dead tissue at their centre; the white areas are alveoli. Magnification ×26.

Bacteria evade the immune system

In the normal course of events, macrophages engulf and destroy bacteria. However, the *M. tuberculosis* bacteria can survive inside macrophages. The bacteria are taken up by phagocytosis, but once inside they resist the killing mechanisms used by these cells. The bacteria have very thick waxy cell walls, making them very difficult to break down. They can lie dormant for years, but if the immune system is weakened the infection can become active again.

Not only can TB bacteria survive and breed inside macrophages, they are also known to target the cells of the immune system. TB bacteria can suppress T cells. This reduces antibody production and attack by killer T cells.

> **Activity**
>
> **Activity 6.9** is a comprehension exercise on TB. **A6.09S**

Active tuberculosis

The second phase (active tuberculosis) occurs if the patient's immune system is unable to contain the disease when it first arrives in the lungs, perhaps because the number of bacteria is too great. Alternatively, an old infection may break out because the immune system is no longer working properly. About 80% of active TB cases are reactivations of previously controlled infections.

The activity of the immune system may be reduced for several reasons. In old age or in the very young (0–5 years) it is less able to respond quickly to pathogens. Malnutrition and poor living conditions also adversely affect the immune system. But the most significant factor in many recent infections is AIDS. HIV, the virus that causes AIDS, directly targets white blood cells and greatly reduces a patient's ability to fight any infection. TB is such an aggressive disease that it kills many people with AIDS, especially in sub-Saharan African countries where millions of people are infected with HIV.

With active tuberculosis in the lungs (known as respiratory pulmonary tuberculosis), the bacteria multiply rapidly and destroy the lung tissue, creating holes or cavities (Figure 6.43). The lung damage will eventually kill the sufferer if they are not treated with an appropriate antibiotic.

Q6.33 What effect will the damage shown in Figure 6.43 have on:
a gas exchange in the lungs
b the breathing rate of the patient?
Explain your answers.

Symptoms of active TB

A patient with active tuberculosis will experience a range of symptoms including:
- coughing – the patient may cough up blood
- shortness of breath
- loss of appetite and weight loss
- fever and extreme fatigue.

The role of fever

A person infected with TB or many other pathogens experiences fever and night
sweats. These occur because, as part of the inflammatory response, fever-causing
substances are released from neutrophils and macrophages. These chemicals
affect the **hypothalamus** and alter the set point for the core body temperature
to a higher temperature. Effectors act to warm the body up to the new set point
(see Key biological principle: Homeostasis and temperature control). The patient
has a high fever, with a temperature in the region of 40.5 °C.

It is thought that the raised temperature enhances immune function and
phagocytosis. In addition, bacteria and viruses may reproduce more slowly
at the high temperatures. TB itself is temperature sensitive and will stop
reproducing at temperatures above 42 °C. However, this temperature
is harmful to the patient. Above 40 °C human enzymes are increasingly
denatured; a fever of 42–3 °C is life threatening.

Q6.34 Why is the denaturing of enzymes life threatening?

Key biological principle: Homeostasis and temperature control

The need for homeostasis
In humans, if cells are to function properly, the body's
internal conditions must be maintained within a
narrow range of the cells' optimum conditions.
The maintenance of a stable internal environment is
called **homeostasis**.

This is partly achieved by maintaining stable conditions
within the blood, which in turn gives rise to the tissue
fluid that bathes the body's cells.

In the blood the concentration of glucose, ions, and
carbon dioxide must be kept within narrow limits.
In addition, the water potential (determined by
the concentration of solutes in the blood), pH and
temperature of the blood are tightly regulated.

The role of negative feedback
Each condition that is controlled has a **norm value** or
set point that the homeostatic mechanisms are

'trying' to maintain. There are **receptors**, which
detect deviations from the norm. These receptors are
connected to a control mechanism, which turns on
or off **effectors** (muscles and glands) to bring the
condition back to the norm value.

For example, as blood glucose concentration rises
above the norm this information is fed back to the
control mechanism. Effectors act to decrease the
blood glucose concentration back towards the norm.
Because a deviation from the norm results in a change
in the opposite direction back to the norm, the process
is known as **negative feedback**. These corrections to
rises above and falls below the norm mean that the
actual value fluctuates in a narrow range around the
norm. This is summarised in Figure 6.44.

(Continued)

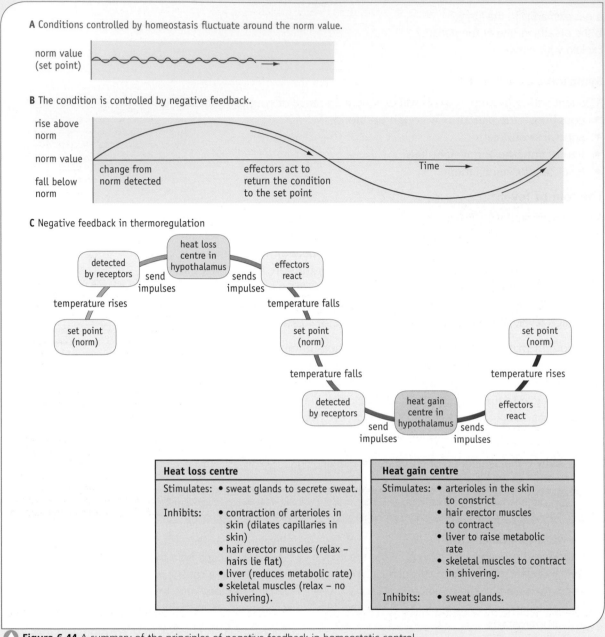

A Conditions controlled by homeostasis fluctuate around the norm value.

norm value
(set point)

B The condition is controlled by negative feedback.

rise above
norm

norm value

fall below
norm

change from
norm detected

effectors act to
return the condition
to the set point

Time

C Negative feedback in thermoregulation

heat loss
centre in
hypothalamus

detected
by receptors

send
impulses

sends
impulses

effectors
react

temperature rises

temperature falls

set point
(norm)

set point
(norm)

set point
(norm)

temperature falls

temperature rises

detected
by receptors

heat gain
centre in
hypothalamus

effectors
react

send
impulses

sends
impulses

Heat loss centre		**Heat gain centre**	
Stimulates:	• sweat glands to secrete sweat.	Stimulates:	• arterioles in the skin to constrict
Inhibits:	• contraction of arterioles in skin (dilates capillaries in skin)		• hair erector muscles to contract
	• hair erector muscles (relax – hairs lie flat)		• liver to raise metabolic rate
	• liver (reduces metabolic rate)		• skeletal muscles to contract in shivering.
	• skeletal muscles (relax – no shivering).	Inhibits:	• sweat glands.

Figure 6.44 A summary of the principles of negative feedback in homeostatic control.

Temperature control

Thermoregulation is the control of body temperature. Our core body temperature is very stable at about 37.5 °C; this is about 0.5 °C higher than oral temperature, measured with a thermometer under the tongue. This body temperature allows enzyme-controlled reactions to occur at a reasonable rate. At lower temperatures the reactions would occur too slowly for the body to remain active; at high temperatures the enzymes would denature.

Temperature control receptors and effectors

In humans, temperature is maintained by a negative feedback system, as summarised in Figure 6.44. This system involves receptors which detect changes in the blood temperature. These receptors are located in the blood vessels supplying blood to a structure in the brain called the hypothalamus. The hypothalamus itself is the control mechanism, receiving information from the receptors and acting as a thermostat, by turning on the effectors necessary to return the temperature to the norm.

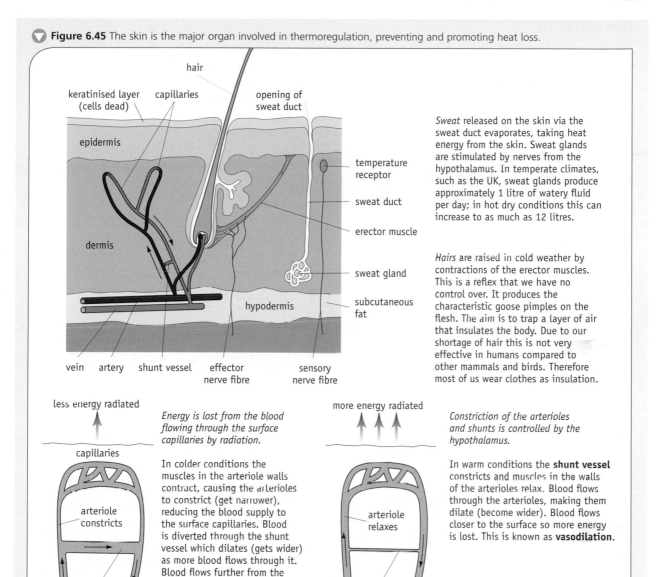

Figure 6.45 The skin is the major organ involved in thermoregulation, preventing and promoting heat loss.

Sweat released on the skin via the sweat duct evaporates, taking heat energy from the skin. Sweat glands are stimulated by nerves from the hypothalamus. In temperate climates, such as the UK, sweat glands produce approximately 1 litre of watery fluid per day; in hot dry conditions this can increase to as much as 12 litres.

Hairs are raised in cold weather by contractions of the erector muscles. This is a reflex that we have no control over. It produces the characteristic goose pimples on the flesh. The aim is to trap a layer of air that insulates the body. Due to our shortage of hair this is not very effective in humans compared to other mammals and birds. Therefore most of us wear clothes as insulation.

Energy is lost from the blood flowing through the surface capillaries by radiation.

In colder conditions the muscles in the arteriole walls contract, causing the arterioles to constrict (get narrower), reducing the blood supply to the surface capillaries. Blood is diverted through the shunt vessel which dilates (gets wider) as more blood flows through it. Blood flows further from the skin surface so less energy is lost. This is known as **vasoconstriction**.

Constriction of the arterioles and shunts is controlled by the hypothalamus.

In warm conditions the **shunt vessel** constricts and muscles in the walls of the arterioles relax. Blood flows through the arterioles, making them dilate (become wider). Blood flows closer to the surface so more energy is lost. This is known as **vasodilation**.

There are also thermoreceptors in the skin that detect temperature changes as a result of environmental changes. If the skin is warm, then messages are sent to the hypothalamus initiating the heat-loss responses and inhibiting heat-gain responses. These changes help keep the core body temperature near its optimum.

There are several ways that the skin can help in the control of temperature, including sweating, hair attitude (flat or raised), route of blood flow, and shivering. The structures in the skin and the methods involved in temperature control are summarised in Figure 6.45. Shivering is the uncontrolled contraction of normally voluntary muscles, which can increase heat production

six-fold. Shivering can transfer energy to muscle tissue to help maintain body temperature.

Q6.35 What effectors are activated when the body temperature falls below the norm?

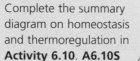

Activity

Complete the summary diagram on homeostasis and thermoregulation in **Activity 6.10**. **A6.10S**

Extension

Find out about homeostasis and control of blood concentration in **Extension 6.1**. **X6.01S**

Glandular TB

Bacteria can also move to infect other parts of the patient's body. Common sites include the bones, lymph nodes and central nervous system. These infections usually follow an initial pulmonary infection, but occasionally these are the only site of disease. In glandular TB the main symptom is enlarged lymph glands, usually in the neck or armpits, although sometimes only lymph glands in the chest are affected which can only be seen on an X-ray. Asian people are more likely to get glandular TB whereas Caucasians get more pulmonary TB.

> ### Did you know? Scrofula – the King's evil
>
> The most common type of glandular TB is a swollen infection of the lymph nodes in the neck (cervical nodes). If you have these swellings you have scrofula.
>
> Scrofula was known as the 'King's evil' because it was believed that the touch of a King could cure the disease. This superstition dates back at least to Edward the Confessor (1002–66). Edward I (1239–1307) touched as many as 533 sufferers in one month but the record must go to Charles II (1630–85) who 'cured' nearly 100 000 people during his reign. References to the royal miracle cure are found in Shakespeare's *Macbeth*.
>
> Before the development of anti-TB drugs in the twentieth century, surgical removal of all the infected lymph nodes was the only alternative treatment. As this would have been done without any real anaesthetic you can see why the royal touch was such an attractive option. There are many written accounts of the time supporting the success of Kings (and Queens) in curing scrofula but the superstition died out in the nineteenth century.

How is TB diagnosed?

To check if someone suspected of having TB really has the disease, a 'history' is taken from the patient. They describe their symptoms to the doctor, which might include feeling generally tired, a persistent cough, weight loss, fever and night sweats. A range of tests can then be used to check for TB.

Skin and blood tests

A skin test may be done (similar to the test you had before your BCG). A positive result shows as an inflamed area of skin around the site of the test injection. Antibodies in the blood cause this inflammation, indicating that TB antigens are already present. Unfortunately the test can give a negative result if the person has latent TB. It can also give a false positive result if the person has had a BCG anti-TB vaccination. To overcome these problems blood tests have been developed to analyse blood samples for T cells specific to antigens that occur only in *Mycobacterium tuberculosis*.

Identification of bacteria

To confirm a positive skin test, a sample of sputum coughed up by the patient is taken and cultured to see what bacteria are present in it. Different bacteria can be identified from the culture by staining techniques. A variety of stains are used, and only some types of bacteria take up particular stains, depending on the make-up of their cell wall.

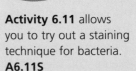

Activity

Activity 6.11 allows you to try out a staining technique for bacteria. **A6.11S**

Chest X-rays

Chest X-rays are usually taken to discover the extent of the damage and disease in the lungs. The patient whose X-ray is shown in Figure 6.46 has extensive infection. X-rays of other organs may also be needed if the disease is thought to have spread outside the lungs.

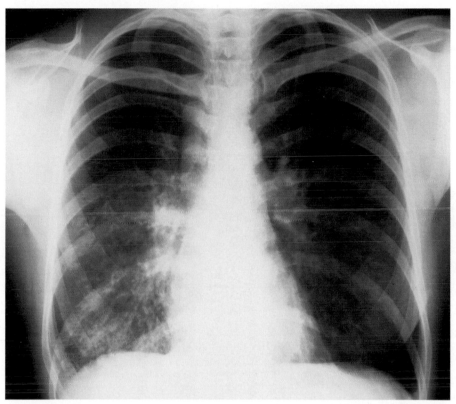

🔺 **Figure 6.46** Chest X-ray of a patient with pulmonary TB. Damaged tissue (red areas) can be seen in the lungs (black).

Put together, these pieces of information will give the doctor a full picture of the condition of the patient. This can then be used to design the best treatment, usually a combination of antibiotics and improved lifestyle.

If the TB sufferer is contagious, all their recent contacts – friends, family, and so on – will need to be tested as well, to check whether they have been infected.

Q6.36 What lifestyle changes might help the patient avoid any future infection with the disease?

6.5 The body's response to HIV and AIDS

What are HIV and AIDS?

Worldwide there is an AIDS epidemic which started in the early 1980s. In 2005, an estimated 40 million people were infected with the virus, with 5 million new infections in 2004. Figure 6.47 shows the very high infection and death rates resulting from HIV/AIDS in sub-Saharan Africa. At the end of 2004, approximately 53 000 people in the UK were living with HIV infection.

Adults and children estimated to be living with HIV/AIDS at the beginning of 2005

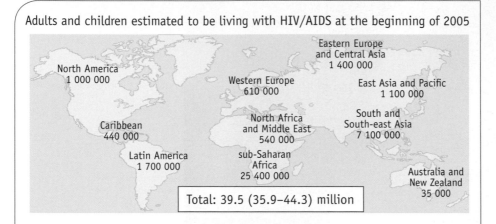

North America
1 000 000

Eastern Europe
and Central Asia
1 400 000

Western Europe
610 000

East Asia and Pacific
1 100 000

Caribbean
440 000

North Africa
and Middle East
540 000

South and
South-east Asia
7 100 000

Latin America
1 700 000

sub-Saharan
Africa
25 400 000

Australia and
New Zealand
35 000

Total: 39.5 (35.9–44.3) million

Estimated number of adults and children newly infected with HIV during 2004

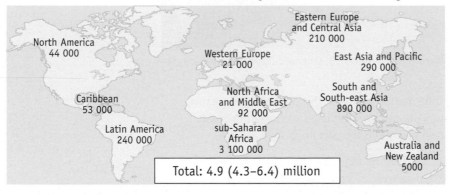

North America
44 000

Eastern Europe
and Central Asia
210 000

Western Europe
21 000

East Asia and Pacific
290 000

Caribbean
53 000

North Africa
and Middle East
92 000

South and
South-east Asia
890 000

Latin America
240 000

sub-Saharan
Africa
3 100 000

Australia and
New Zealand
5000

Total: 4.9 (4.3–6.4) million

Estimated adult and child deaths due to AIDS during 2004

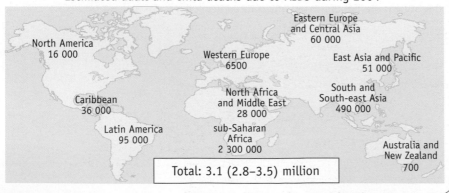

North America
16 000

Eastern Europe
and Central Asia
60 000

Western Europe
6500

East Asia and Pacific
51 000

Caribbean
36 000

North Africa
and Middle East
28 000

South and
South-east Asia
490 000

Latin America
95 000

sub-Saharan
Africa
2 300 000

Australia and
New Zealand
700

Total: 3.1 (2.8–3.5) million

Figure 6.47 HIV infection is not distributed evenly across the globe. The figures are only estimates; the possible range of values is indicated alongside each total figure. *Source: UNAIDS.*

AIDS, **acquired immune deficiency syndrome**, is caused by infection with the human immunodeficiency virus, HIV (Figure 6.48). A syndrome is a collection of symptoms related to the same cause, in this case the action of HIV which gradually destroys part of the immune system. The symptoms of AIDS are those of opportunistic infections to which the patient becomes susceptible as their immune system is weakened.

Weblink

You can find out the latest figures by visiting the UNAIDS or World Health Organisation websites.

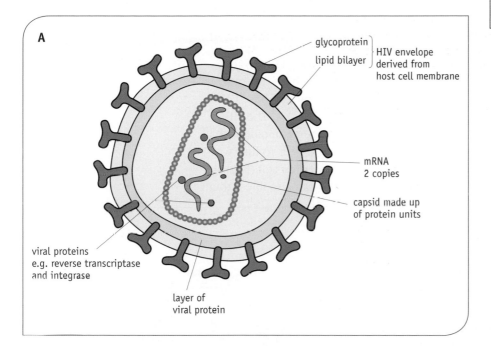

A

glycoprotein
lipid bilayer
} HIV envelope derived from host cell membrane

mRNA 2 copies

capsid made up of protein units

viral proteins e.g. reverse transcriptase and integrase

layer of viral protein

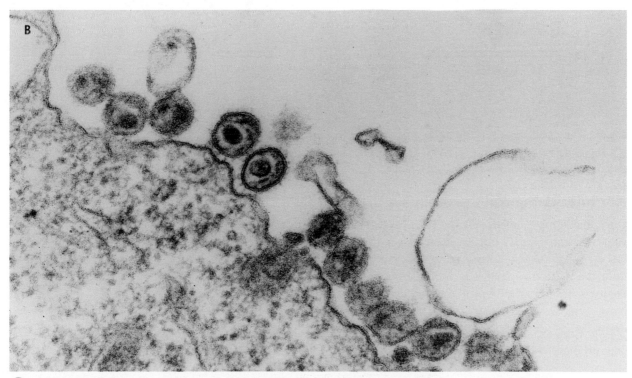

B

▲ **Figure 6.48 A** The structure of the human immunodeficiency virus. **B** The transmission electron micrograph shows the surface of a T cell with HIV particles. Magnification ×120 000.

In Figure 6.48 you can see the internal structure of HIV. HIV is a structurally complex virus, an example of an enveloped virus. The lipid envelope is formed from the host cell membrane as the new virus particles emerge from the cell cytoplasm. Sticking through the envelope are viral glycoprotein (gp) molecules.

Activity

Activity 6.12 looks at the structure of the HIV virus in detail. **A6.12S**

Infection with HIV

HIV invade T helper cells

Particular glycoprotein molecules called gp120 on the virus surface are important for binding the virus to the surface of the host cells. HIV invades T helper cells within the immune system. The T helper cells have receptor molecules on their cell surface membrane called CD4 receptors. These CD4 receptors are involved in the immune response (see Figure 6.38). The HIV gp120 molecules also attach to the CD4 receptors. They then combine with a second receptor. This allows the envelope surrounding the virus to fuse with the T helper cell membrane, enabling the viral RNA to enter the cell (Figure 6.49). Macrophages also have CD4 receptors, so the virus can also infect them.

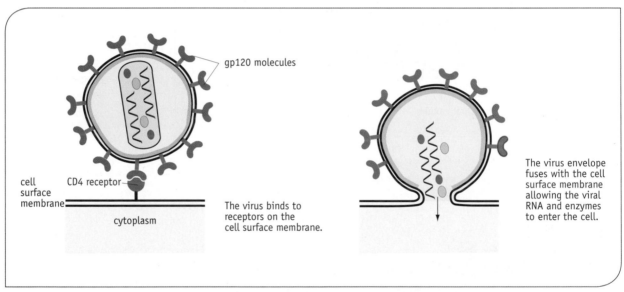

Figure 6.49 The gp120 spikes on the surface of the HIV attach to receptors on the cell surface, allowing the virus envelope to fuse with the host cell membrane.

HIV hijacks the cell's protein synthesis

Once inside the T helper cell, the virus needs to make the cell replicate new viral components. HIV nuclear material is in the form of RNA not DNA, so the first step is to reverse normal transcription and manufacture DNA from the RNA template. To do this, the virus uses an enzyme called **reverse transcriptase**. Figure 6.50 shows the steps in this process. Viruses that contain RNA and use reverse transcriptase in this way are known as retroviruses.

Q6.37 Copy Figure 6.50 and recall the principles of base pairing to complete the gaps on the DNA.

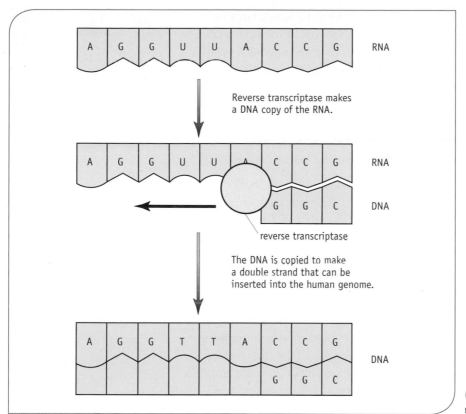

Figure 6.50 The action of reverse transcriptase.

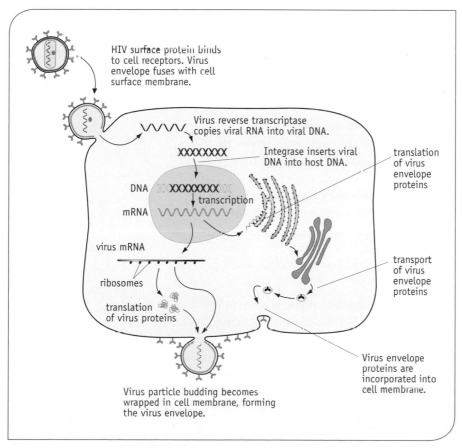

Figure 6.51 A simplified view of the HIV life cycle.

Once the DNA strand is produced, it is integrated into the host's DNA by another HIV enzyme, **integrase**. Now, like any other part of the host's DNA, the viral DNA can be transcribed and translated to produce new viral proteins (Figure 6.51). These new viral proteins, together with glycoproteins and nuclear material, are assembled into new viruses.

T helper cells destroyed

These new viruses bud out of the T cell, taking some of the surface membrane with them as their envelope, and killing the cell as they leave.

Infected T helper cells will also be destroyed by T killer cells. Thus as the number of viruses increases, the number of host T helper cells decreases. The loss of T helper cells results in macrophages, B cells and T killer cells not being successfully activated and therefore not functioning properly. This means that the infected person's immune system becomes deficient.

The course of the disease

You probably know that AIDS does not always follow HIV infection straight away. There are several stages in the course of the disease once someone is infected, and this depends on many factors. The health of the host before infection, their genetic resistance to infection, the quality of immune response to the initial infection, and the availability of the currently expensive and complicated drug treatment all contribute to an individual's disease and life expectancy. There is currently, and for the foreseeable future, no cure for AIDS. AIDS nearly always follows HIV infection, eventually, and has 100% mortality unless treated by a cocktail of drugs.

The acute phase

When a person is first infected by HIV, there is an acute phase of infection. The following events occur during this phase:
- HIV antibodies appear in the blood after 3–12 weeks.
- The infected person may experience symptoms such as fever, sweats, headache, sore throat and swollen lymph nodes, or they may have no symptoms.
- There is rapid replication of the virus and loss of T helper cells.
- After a few weeks, infected T helper cells are recognised by T killer cells which start to destroy them. This greatly reduces the rate of virus replication but does not totally eliminate it.

The chronic phase

There is now a prolonged chronic phase. The chronic phase is sometimes called the 'latent' phase. However this implies that the virus is not active, which is not the case. The virus continues to reproduce rapidly, but the numbers are kept in check by the immune system. The phase is characterised by the following features:
- There may be no symptoms during this phase, but there can be an increasing tendency to suffer colds or other infections which are slow to go away.
- Dormant diseases like TB and shingles can reactivate.

The chronic phase can last for many years. In fit young people with a healthy lifestyle it can last for 20 years or more, especially if combined with drug treatment. Unfortunately, most HIV infection occurs in developing countries with little money available to provide drugs, and amongst people who often do not have access to sufficient food or clean water. These people may go on to develop AIDS within a few years of infection.

The disease phase

Eventually, the increased number of viruses in circulation (the viral load) and a declining number of T helper cells indicates the onset of AIDS, the disease phase. During this phase, the decrease in number of T helper cells leaves the immune system vulnerable to other diseases. A normal T helper cell count is over 500 per mm^3 of blood; below 200 per mm^3 of blood there is a high risk of infection by diseases that take advantage of the weakened immune system – so-called **opportunistic infections** which can rapidly be fatal.

Opportunistic infections often found in people with AIDS include pneumonia and TB. There can be significant weight loss. Patients can develop dementia (memory and intellect loss). People with AIDS are also susceptible to tumours such as Kaposi's sarcoma. This is rare in the general population, and it was the sudden occurrence of several cases of people with Kaposi's sarcoma, noticed by doctors in the USA in 1981, which contributed to the identification of AIDS as a new disease. This cancer is readily identified by purple-black patches on the skin.

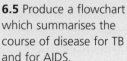

Checkpoint

6.5 Produce a flowchart which summarises the course of disease for TB and for AIDS.

Did you know? Where diseases come from

The common diseases that kill or have killed huge numbers of people, such as smallpox, TB, malaria, flu, cholera and AIDS, have evolved from diseases of animals that have been transferred to humans. Molecular biologists can identify the closest relatives of these disease microbes. For instance, measles evolved from rinderpest, a disease of cattle, TB from *Mycobacterium bovis* in cattle, and HIV from a virus found in wild monkeys.

Close contact between livestock and humans increases the chances of transfer from animals to humans. We can occasionally pick up infections from wild or domestic animals, such as leptospirosis from dogs and brucellosis from cattle. These pathogens are at an early stage of evolution, and cannot yet spread between human hosts, unlike more specialised human pathogens.

Many epidemic diseases evolved and became widespread in Europe when people lived in close contact with their animals. The diseases were then taken to the New World, where the native people of the Americas fell victim to the new diseases. The conquerors of the Aztecs and Native Americans used violence, but their microbes were the real conquerors.

6.6 Could the infections have been prevented?

Tests showed that Nicki Overton had a high HIV viral load and low T helper cell count. This suggests that she had entered the disease phase of the infection, and she would have been very susceptible to opportunistic infections. She had been staying in hostels while travelling, and the crowded conditions could have exposed her to TB bacteria. Her weakened immune system would have been unable to deal with TB, which resulted in her death. Could the body have prevented entry of the pathogens? Could she have developed any immunity to the diseases? Could she have been vaccinated against the diseases or given treatment? Find out below.

Preventing entry of pathogens

Normally the body has mechanisms to prevent the entry of pathogens. These include physical barriers and chemical defences.

The skin

The skin's keratin (hard protein) outer layer is effective in stopping entry of microorganisms. Entry can occur through any wounds, but blood clotting seals the wound and thus reduces the number of microorganisms gaining access. Large numbers of microbes (known as the **skin flora**) live on the skin surface. These are harmless and prevent colonisation by other bacteria. The bacteria that occur naturally on the skin surface are well adapted to the environment on the skin. Other bacteria are not so well suited to the conditions created by salty sweat and excreted chemicals such as urea and fatty acids.

Mucous membranes

The mucous membranes that line the airways and gut provide easier routes into the body because, lacking any keratin layer, the surface is always moist, making it a more favourable environment for bacterial growth. As described in Topic 2, entry of microbes to the lungs is limited by the action of **mucus** and **cilia**. The mucus, secreted by goblet cells in the trachea and bronchi, traps microbes and other particles; then beating cilia carry the mucus up to the throat where it is swallowed. Secretions in the mouth, eyes and nose contain **lysozyme**, an enzyme that breaks down bacterial cell walls, causing the cell to burst.

Digestive system

Stomach acid

Acid in the stomach kills most bacteria that enter with food. Gastric juices secreted by gastric glands in the stomach wall contain hydrochloric acid giving a pH of less than 2.0, the optimum pH for the enzyme pepsin that is also secreted in the gastric juices.

Gut flora

Bacteria are found in the small and large intestines. There are hundreds of different species found within the intestines including harmless strains of *Escherichia coli*. These natural flora benefit from living within the gut where conditions are ideal: warm, moist and with plentiful food supplies. The host (each of us) benefits from their presence; it is a mutualistic relationship. The bacteria may aid the digestive process, and competitively exclude pathogenic bacteria, competing with the pathogens for food and space. The bacteria also secrete chemicals such as lactic acids which are useful in the defence against pathogens.

Some foods contain living bacteria to help improve the microbial balance in the large intestine. The bacteria used in these probiotic foods are usually the lactic-acid-secreting bacteria.

Becoming immune

You can become immune to a specific disease in various different ways. Use the profiles in Table 6.9 to find out the differences between:

- **passive natural immunity**
- **active natural immunity**
- **active artificial immunity**
- **passive artificial immunity**.

Activity

Complete the summary diagram in **Activity 6.13** showing the mechanisms that prevent microbe entry. **A6.13S**

Extension

Read about the development of probiotic foods, foods that contain useful bacteria, in **Extension 6.2**. **X6.02S**

Table 6.9 Natural and artificial immunity.

Type of immunity	Example
passive natural immunity	Catherine has just been born. Her immune system is undeveloped but antibodies that have been crossing the placenta from her mother during the past three months protect her.
passive natural immunity	Simon is one month old and his immune system is not fully developed. He is protected from infectious diseases by antibodies received from his mother's milk. The colostrum, or first-formed milk, which he had during his first few days was particularly rich in maternal antibodies. In addition he is still getting protection from the maternal antibodies that crossed the placenta before he was born.
active natural immunity	Owen is three years old and is recovering from chickenpox. His body's specific immune response to the foreign antigens helps to destroy the virus and make him immune to chickenpox in the future. He has a supply of antibodies and B memory and T memory cells in his system that will respond quickly if he is reinfected with the pathogen.
active artificial immunity	Sital is about to start school. She has been vaccinated against diphtheria, tetanus, whooping cough, meningitis C, polio, measles, mumps and rubella. The antigens in the vaccines did not cause the diseases but in each case stimulated a specific immune response. This has given her immunity to all eight of these serious and potentially deadly illnesses.
active artificial immunity	Students in class 9K are being vaccinated against meningitis C. This vaccine was introduced in 1999, initially for young people. It was hugely successful and by 2001 there had been a 90% reduction in the UK in both meningitis C cases and deaths. In 2002 the vaccination programme was extended to all 20–24-year-olds. The vaccine does not give immunity to all types of meningitis and anyone showing symptoms, even if they have been vaccinated, should still get urgent medical treatment.
passive artificial immunity	Chris got a long thorn in her finger while she was gardening and is at risk from tetanus bacteria which may be in the wound. Because she has not been vaccinated against tetanus she needs immediate protection. She is therefore being given an injection of tetanus antibodies that should stop her getting the disease.

Q6.38 Which types of immunity:
a give immediate protection
b develop after a time lag
c last for a short time, perhaps only a few weeks
d give long-lasting protection
e involve memory cells
f require medical treatment?

Being vaccinated

Gaining immunity

When you are vaccinated for a particular disease, your immune system responds to the **vaccine** in the same way as it responds to the disease. Antibodies are produced and memory cells ensure lasting protection. In future you can rapidly destroy the pathogen before the onset of any symptoms – you are said to be immune.

A vaccine must contain one or more antigens that are also found on the pathogen or its toxin. This can be achieved in several different ways. Vaccines may contain:

- attenuated viruses. Attenuated viruses have been weakened so that they are harmless. For example, the measles vaccine contains attenuated measles viruses.
- killed bacteria. One of the commonly used whooping cough vaccines contains bacteria capable of causing whooping cough which have been killed.
- toxin that has been altered into a harmless form. The diphtheria vaccine contains a harmless form of the toxin produced by *Corynebacterium diphtheriae*, a bacterium that causes diphtheria.
- an antigen-bearing fragment of the pathogen. Some of the newer vaccines, like that for meningitis C, are of this type.

After the initial vaccination it is usually necessary to have one or more 'boosters' to ensure long-lasting immunity. In the case of influenza, vaccination with a completely new vaccine is required every year in order to protect against new strains of the virus.

Vaccination for infectious diseases protects not just the individual but also the community. When enough people are immunised, the disease is less likely to be transferred from one person to another and so there is less disease in the community as a whole. This means that anyone who did not respond to the vaccine or could not be given it for medical reasons is still, in effect, protected. This group protection is called **herd immunity**. For measles, which is extremely infectious, 95% of the population need to be vaccinated to achieve herd immunity.

Q6.39 **a** Use Figure 6.52 to find out the year in which herd immunity to measles was first achieved in England and Wales.
b Suggest why, before 1968, there was a measles epidemic every two years.
c Why were there still epidemics between 1968 and 1988?

Checkpoint

6.6 Write a definition for each of the four types of immunity:
passive natural immunity, active natural immunity, active artificial immunity and passive artificial immunity.

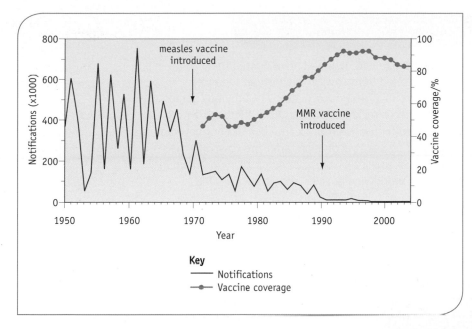

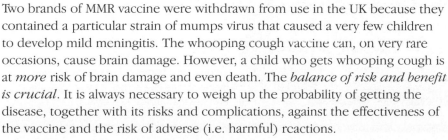
Figure 6.52 Annual measles notifications and vaccine coverage in England and Wales, 1950–2003. *Source: Office for National Statistics and Health Protection Agency.*

There is currently no vaccine against AIDS, but children in British schools are routinely vaccinated against TB. The BCG vaccine is given to young people between the ages of 10 and 13 years. Before the vaccination is given, each child is first given a test to determine if they have natural immunity from a previous infection. An extract from TB bacteria is injected just below the skin using a ring of small needles. If the area of skin becomes red and swollen after about three days, the test is positive (Figure 6.53), and the individual has immunity to the disease. About 8% of UK children test positive.

If the test is negative, a BCG vaccination is given. This contains live, chemically attenuated bacteria. (BCG stands for bacille (bacteria) of Calmette and Guérin, after the two French doctors who developed the vaccine.)

Are vaccinations dangerous?

Some vaccinations cause mild soreness at the site of the injection, fever or a general feeling of being unwell.

Research studies have indicated that, very occasionally, much more serious and long-term damage can be linked to certain vaccines, some of which have now been withdrawn.

Figure 6.53 A positive result for a heaf test, the BCG test injection. The red swelling appears within about three days.

Two brands of MMR vaccine were withdrawn from use in the UK because they contained a particular strain of mumps virus that caused a very few children to develop mild meningitis. The whooping cough vaccine can, on very rare occasions, cause brain damage. However, a child who gets whooping cough is at *more* risk of brain damage and even death. The *balance of risk and benefit is crucial*. It is always necessary to weigh up the probability of getting the disease, together with its risks and complications, against the effectiveness of the vaccine and the risk of adverse (i.e. harmful) reactions.

Edward Jenner (1749–1823), British physician and naturalist, investigated the observation that milkmaids who had suffered from cowpox did not get the more serious and often fatal infection smallpox. In 1796 he inoculated a healthy boy, James Phipps, with fluid from a cowpox blister on a milkmaid's finger. After the boy had recovered from cowpox Jenner inoculated him with smallpox. The boy did not develop smallpox. Jenner named the immunising process 'vaccination' after the cowpox virus (vaccinia, Figure 6.54). Vaccination against smallpox was made compulsory in Britain in 1853 and smallpox was declared extinct worldwide outside laboratories in 1980.

Long before Jenner came up with the idea of vaccination in 1796, people in the Middle East, Africa and Asia were inoculating against smallpox. Lady Mary Wortley Montague, the wife of the British Ambassador in Constantinople, wrote in 1717 describing the process:

> *The small pox, so fatal, and so general amongst us, is here entirely harmless, by the invention of engrafting ... the old woman comes with a nutshell of the matter (pus) of the best sort of small pox, and asks which vein you chose to have opened. She immediately rips open that you offer her, with a large needle (which gives you no more pain than a common scratch) and puts into the vein as much matter as can lie on the head of her needle ... Every year, thousands undergo this operation ... There is no example of any one that has died in it.*

Lady Mary was so impressed that she had her son inoculated in Constantinople and, on her return to England, tried hard to spread the practice in Britain. She did manage to convince the King and Queen. After an ethically questionable set of tests on prisoners and orphans they had their own children inoculated in 1721. Despite royal approval inoculation never became very widespread in Europe. This may have been because those who had been inoculated sometimes developed full-blown smallpox.

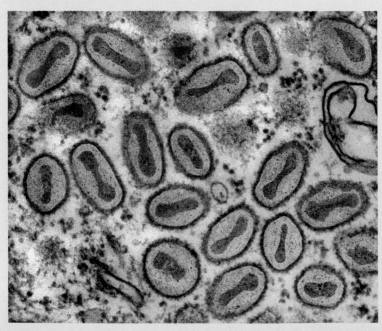

◀ **Figure 6.54** Coloured transmission electron micrograph (TEM) of sectioned vaccinia virus particles. The virus is covered by membrane layers (green) taken from the host cell that replicated the virus.

6.7 Are there treatments for AIDS and TB?

AIDS

There is no treatment to get rid of HIV in someone who is infected with the virus because the virus is hidden inside the T helper cells. But there are drugs available which reduce the production of more viruses. There are two main types of drug:

- reverse transcriptase inhibitors, which prevent the viral RNA from making DNA for integration into the host's genome
- protease inhibitors, which inhibit the proteases that catalyse the cutting of larger proteins into small polypeptides for use in the construction of new viruses.

Other types of drug are undergoing clinical trials. HIV can develop resistance to anti-HIV drugs and therefore these drugs are often given in combination. If the virus becomes resistant to one drug, it may be still be susceptible to the other drugs being taken. Unfortunately, because the anti-HIV drug treatments are very expensive and need to be taken for extended periods of time, they are not available to the vast majority of people who are infected.

Treating tuberculosis

Active TB bacteria can be killed by antibiotics. Usually a combination of four antibiotic drugs is given for two months, with two drugs continued for a further four months to ensure that any dormant bacteria are destroyed.

What are antibiotics?

More than 3000 years ago, the Egyptians, the Chinese and Central American Indians used moulds to treat rashes and infected wounds. They did not understand what caused the diseases or how the mould helped to treat them; in some cases they believed that the moulds drove away evil spirits that caused the disease.

In 1928, a research scientist by the name of Alexander Fleming was working in a hospital in London. He was studying the bacterium *Staphylococcus aureus*. One day Fleming found a spot of green growing in one of the agar plates. He noticed there was a clear, bacteria-free ring of agar gel around the mould. He thought this might mean that the mould had killed the bacteria. Fleming watched the dish over the next couple of days and saw that the more the mould spread, the more of the bacteria that were killed. He noticed that the mould produced tiny droplets of fluid on its surface (Figure 6.55), and he wondered if this was the chemical that was destroying the bacteria.

Fleming drew off the liquid. He found that this liquid could kill bacteria in a test tube. The name of the mould was *Penicillium notatum*, so he decided to call the liquid penicillin. Later, other scientists discovered that penicillin could cure certain infections in mice and rabbits without harming the animals in any way. Fleming was working on the development of vaccines, so he did not take his discovery any further.

◀ **Figure 6.55** The fungus *Penicillium notatum* growing on an agar plate.

Q6.40 One explanation for the effects of penicillin was that it is a toxic chemical that kills bacteria. Another was that it is an enzyme that digests the bacteria. Describe how you could carry out an investigation to distinguish between these two hypotheses.

Searching for new antibiotics

Penicillin had been proven to work against pneumonia, scarlet fever and several other diseases. However, it had no effect on the bacteria that caused typhoid, TB or many other diseases.

In 1932, an American called Selman Waksman showed that when *Mycobacterium tuberculosis* was added to certain types of soil the bacteria died. In 1943 Waksman and a research student called Albert Schatz tested hundreds of soil organisms for antibacterial activity. After three months they isolated an organism called *Streptomyces griseus* from a sick chicken. They found it produced an antibiotic called streptomycin, which they purified. Streptomycin was the first drug found that could cure tuberculosis.

In 1953, Waksman decided to call these antibacterial chemicals antibiotics, and he defined the term as '... a chemical substance, produced by microorganisms, which has the capacity to inhibit the growth and even to destroy bacteria and other microorganisms, in dilute solutions'. In 1952 Selman Waksman was awarded a Nobel Prize for the discovery of streptomycin.

Q6.41 Imagine that you are a scientist working in the 1940s. You have a limited amount of streptomycin and you have been asked to carry out a test to see whether it is an effective treatment for tuberculosis, caused by *Mycobacterium tuberculosis*.

You have three possible options:

1 Treat the next 100 patients with TB coming into the hospital, and compare them with previous TB patients.

2 Give streptomycin to every second TB patient who comes into the hospital, and compare them with the untreated ones.

3 Select TB patients at random, and give them the streptomycin. Compare their progress with the untreated patients.

Which is the best way of testing how effective streptomycin is at treating TB?

Penicillin and streptomycin were each effective against certain diseases, but scientists wanted a broad-spectrum antibiotic that could cure many different diseases. During the 1950s and 1960s a large number of new antibiotics were developed. However, even with the discovery of a large range of antibiotics there are still bacterial infections that don't respond to them, or become resistant to them. The search goes on.

Why do some microorganisms make antibiotics? The usual answer is that antibiotics help microorganisms to compete in the environment. For example, a soil-living fungus may secrete an antibiotic so that other fungi or bacteria are unable to grow near it. However, some scientists are not convinced that microorganisms make antibiotics for this reason. Antibiotics are not produced in large amounts until the cells are ageing. If antibiotics were produced to help a microorganism compete, it would be expected that they would be produced mainly in young cells.

One feature of antibiotics is that they are effective against bacterial cells but leave eukaryotic cells unharmed. They are also useless against viruses, which is why diseases such as colds and flu should not be treated with antibiotics.

Did you know? Penicillin production

In 1939 a scientific team lead by Howard Florey and Ernst Chain was set up in Oxford to produce usable quantities of penicillin. Ernst Chain developed a freeze-drying technique for purifying penicillin, but mass production was difficult. They started to grow *Penicillium notatum* in containers in their labs. They used this penicillin for the first clinical trials, but they could not produce enough.

Howard Florey knew that penicillin would be useful to treat infected wounds in soldiers during the Second World War, but British chemical companies were producing explosives and were unable to mass-produce penicillin as well. When the USA joined the war in 1941, American chemical companies started to produce the drug. However, it was not until 1944 that there was enough penicillin to meet all the needs of the allies.

The Nobel Prize for Medicine in 1945 was awarded to Fleming, Florey and Chain for the discovery and production of penicillin.

Penicillin is now produced on a large scale. It can be made by microbial fermentation, or by a semi-synthetic process in which one antibiotic is converted into another. The fungus *Penicillium chrysogenum* is grown in stainless steel fermenters that contain up to 200 000 dm^3 of nutrient medium (Figure 6.56). The penicillin is released into the fluid. By adjusting the pH at the end of the fermentation process, the penicillin crystallises and can be extracted. Then, using immobilised enzyme technology (see Activity 1.16), the penicillin can be modified to give a range of novel penicillins.

▲ **Figure 6.56** Large-scale fermenters being used to produce penicillin.

How antibiotics work

Classifying antibiotics

Antibiotics are classified according to their method of action. There are two types:

- **Bactericidal** antibiotics destroy bacteria.
- **Bacteriostatic** antibiotics prevent the multiplication of pathogens. The host's own immune system can then destroy the pathogens.

Q6.42 Clear zones around a test disc impregnated with various antibiotics provide a method of comparing the effectiveness of different antibiotics (Figure 6.57). This method can also be used to compare disinfectants. The wider the zone that remains clear, due to no bacterial growth, the more effective the antibiotic is at preventing growth. Look at the plate below and decide which of the six antibiotics being tested is the most effective.

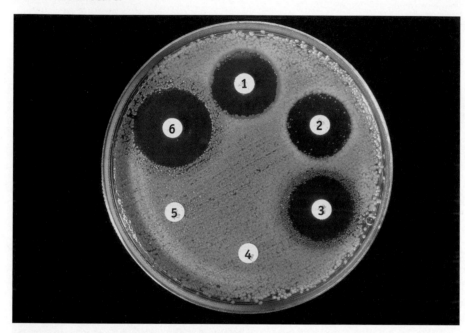

◀ **Figure 6.57** Each disc contains a different antibiotic.

How antibiotics disrupt bacterial cell growth and division

There are several ways in which antibiotics can interfere with bacterial cell growth and division. These include the following:

- Inhibition of bacterial cell wall synthesis. If a weak cell wall forms this can lead to lysis (bursting) of the cell.
- Disruption of the cell membrane, causing changes in permeability that lead to cell lysis.
- Inhibition of nucleic acid synthesis, replication and transcription. This prevents cell division and/or synthesis of enzymes.
- Inhibition of protein synthesis: essential proteins are not produced.
- Inhibition of specific enzymes found in the bacterial cell but not in the host.

Q6.43 The antibiotic vancomycin blocks bacterial cell wall synthesis. Why would it not affect the cells of a person taking the antibiotic?

Activities

In **Activity 6.14** you can compare the effectiveness of different antibiotics. **A6.14S**

Activity 6.15 will test your understanding of bactericidal and bacteriostatic antibiotics. **A6.15S**

Strategies for finding new antibiotics

A variety of approaches are employed in the development of new antibiotics, including detailed study of the structure and function of the pathogens.

One of the ways in which bacteria are able to evolve antibiotic resistance is by pumping antibiotics out of themselves using a cell membrane protein. Scientists are now studying how these pumps work, hoping that this will enable them to develop new antibiotics that either block these pumps or are resistant to their effects. Researchers are working hard to determine the three-dimensional shape of these protein pumps. These proteins are found only in bacteria, so this makes them ideal targets for drugs.

Q6.44 **a** Explain why scientists need to work out the three-dimensional shape of the proteins involved in pumping antibiotics out of the cell.
b Why is it useful that these proteins are found only in bacteria and not in humans?

Another important line of research is sequencing the genomes of pathogenic bacteria. Scientists can then look at which genes are working and which proteins are being produced when the bacterium is in different environments. They can tell how much a gene is being expressed by measurement of the messenger RNA it produces. Scientists can also investigate how introducing mutations into specific genes alters the function of the bacterium. It would be very useful if they could identify genes involved in attaching to host cells or entering them, so that they could then develop strategies for blocking these genes.

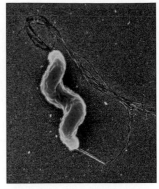

△ **Figure 6.58** Scientists working on this bacterium, *Campylobacter*, which causes serious food poisoning, have found that it has a poor DNA repair system, so that the genes coding for its surface proteins mutate frequently. This has given the bacterium an effective way of evading the host's immune system.

Why do we still have diseases like TB?

An evolutionary race

Disease has been a feature of human life since we first evolved. Pathogens and hosts have always lived with each other. In the millennia of coexistence, the selective pressure exerted by the pathogens has resulted in selection for mutations in the human genome to make us more resistant. As we saw in Topic 2, carriers of the sickle cell allele are more resistant to malaria, whilst the CF allele may provide resistance against cholera. The Tay-Sachs gene that causes progressive degeneration of brain function is common among Ashkenazi Jews, but carriers are more resistant to TB.

So why are we still dying of diseases that have been around for centuries? The problem is that the pathogens and host are locked in an evolutionary arms race. As quickly as we evolve mechanisms to combat pathogens, they evolve new methods of overcoming our immune system. Sometimes pathogens have the edge in this race because their reproductive cycle is faster than ours and their population size is greater.

During the time that TB has been infecting humans the selection pressure exerted by the action of the immune response has resulted in evolution of the bacteria.

Bacterial populations evolve very quickly. There are several reasons for this.

1 Bacteria reproduce very fast. For example, one *E. coli* bacterium can divide every 20 minutes, producing 2 million new cells in just two hours. These cells contain a total of 8000 million genes which will, on average, include a total of 800 mutations.

Activity

Use **Activity 6.16** to link TB and some other ideas covered in the topic.
A6.16S

2 Bacterial population sizes are usually in the billions, so the number of cells containing mutations is vast.

3 Some of these mutations will be advantageous to the cell containing them. They may allow the cell to use different food resources, reproduce more quickly, infect other cells more successfully, or produce symptoms in the host, such as coughing and sneezing, which aid the spread of the disease. Bacteria with a useful mutation are more likely to reproduce and spread.

Since the human immune system is one of the main selection pressures acting on bacteria which infect humans, it is not surprising that there has been an evolution of bacterial strategies for evading or disabling the immune system. For example, slight changes in the pathogen's antigens mean that any reservoir of antibodies, and B and T memory cells from a previous infection, will be useless in combating a second infection.

We are not winning the race

It might appear that the discovery of penicillin and the development of an array of other antibiotics has given us victory in our race with the pathogens. It was thought that the battle against disease would be won with the use of these wonder drugs, and for a time it seemed as if it would be. Eradication of TB in Britain by the late 1970s was considered possible. But over 30 years later it remains a major killer, and we are far from eradicating it.

Antibiotics provide another selection pressure

Mutations arise in pathogenic bacteria that can make them resistant to antibiotics. The bacteria may produce an enzyme that enables the cell to break down the antibiotic, or use a different metabolic pathway for the reactions inhibited by the antibiotic.

In the absence of the antibiotic, bacteria with the mutation may be at a disadvantage. They may reproduce more slowly, using resources to produce enzymes that under 'normal' circumstances (no antibiotic present) are not required.

However, the presence of the antibiotic produces a selection pressure. Those bacteria that do not possess the gene for resistance are selected against. Those that have the gene are selected for; they survive, grow and reproduce. The frequency of the gene within the bacterial population will increase. Since the advantageous gene is passed vertically from one generation to the next, this is sometimes termed vertical evolution.

In bacteria there is also horizontal evolution when the gene is passed from one bacterium to another, which may be the same or a different bacterial species. Bacteria do not undergo the sort of sexual reproduction that most animals do, but they do have cell-to-cell contact in a process called **conjugation** (Figure 6.59).

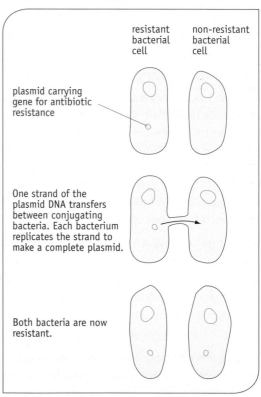

▲ **Figure 6.59** During conjugation, a copy of a DNA plasmid can be transferred between the cells. Antibiotic-resistance genes are usually carried on these small loops of DNA, so when the plasmid is passed on the resistance is transferred.

Antibiotic-resistant TB

The antibiotic streptomycin was the major weapon in the treatment of TB. However, there is now widespread resistance to this and other antibiotics. In the presence of the antibiotic any resistant bacteria will be at an advantage, because they survive, and with less competition reproduce rapidly. This explains why some people with the disease may appear to improve at first in response to a single drug, and then worsen as the drug-resistant mutants multiply unchecked. A different antibiotic is then needed to clear up the infection. For many years this phenomenon did not present a problem, as new antibiotics were being developed faster than the bacteria developed resistance. In recent years, though, the bacteria seem to be catching up, with fewer new drugs being successfully developed.

Pathogens versus the drug developers

Some bacteria have evolved resistance to several antibiotics. These multiple-resistant strains are creating particular problems for the treatment of infected patients. Ironically these bacteria are common in hospitals. For example, *Staphylococcus aureus*, a bacterium that normally causes few problems, has become resistant to most antibiotics including methicillin. This strain, known as methicillin-resistant *Staphylococcus aureus* (MRSA) can cause a dangerous infection. This highly resistant bacterium can be controlled using the antibiotic vancomycin, but there are cases of vancomycin resistance and the bacterium is continually evolving resistance to any new drugs. For example, when a new drug linezolid was introduced in January 2001, some MRSA bacteria had developed resistance to it by December 2002 even though in an attempt to prevent bacteria from evolving resistance the drug had not been extensively used. An evolutionary race exists between pathogens and drug developers.

Multiple-resistant TB is still relatively rare in the UK, but internationally there is growing concern about its spread. Preventing the development and spread of multiple-resistant bacteria is helped by a variety of methods:

- Antibiotics should only be used when needed.
- Patients should complete their treatment even when they feel better, so that all the bacteria are destroyed.
- Infection control should be used in hospitals to prevent spread.

However, David Livermore, director of antibiotic-resistance monitoring at the UK Public Health Laboratory Service, has said: 'The development of new antibiotics is essential because however well we use antibiotics and however good (we are) at stopping the spread of infection, I don't honestly believe we can beat evolution.'

Q6.45 Nicki Overton was infected with HIV and *M. tuberculosis*. What precautions could she have taken to avoid becoming infected?

Activity

Use **Activity 6.17** to check your notes using the topic summary provided. **A6.17S**

Topic test

Now that you have finished Topic 6, complete the end-of-topic test before starting Topic 7.

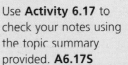

Topic ⑦ Run for your life

Why a topic called Run for your life?

Figure 7.1 The fastest land mammal, the cheetah, at full stretch.

Cheetahs can run at speeds in excess of 100 km/h but after just a few hundred metres they must stop and rest or risk collapsing. Wildebeest can run, though not as quickly, for many kilometres. Both animals are literally running for their lives. Whether chasing prey or seeking out new grazing pasture, their survival depends on their ability to run.

Before humans began farming just 10 000–12 000 years ago we all lived as hunter-gatherers, working from a temporary home base, hunting wild animals and gathering plants from our natural surroundings. There would be long periods of moderate exercise while plucking shoots and berries, with occasional vigorous activity such as chasing after prey with spears.

Figure 7.2 Wildebeest travel huge distances during their annual migrations to find fresh pasture.

Although today few humans have to chase down prey or cover huge distances on foot, many of us still run. Exercise that involves running, or at least jogging, helps maintain our health, and for the professional sportsperson it also provides a living. We marvel at those who, like Paula Radcliffe in Figure 7.4, can complete a 43 km (26 mile) marathon in around two and a quarter hours and are truly amazed by the ultra-marathon runners who cover 100 km, running for as long as 10 hours. But how is it that the marathon runner who can complete 43 km at a pace of about 20 km/h would fail to run 100 metres in the 10 seconds that it takes a top-class sprinter?

The cheetah, wildebeest, humans and all other mammals share the same basic structures that allow us to move around. Our bones, joints and muscles are very similar, both macroscopically and microscopically, and all mammals, indeed most animals, use the same biochemical pathways to make energy available for movement. But if we are all so similar, how is that the wildebeest and caribou can keep going for hours on end but the cheetah must rest after just a few minutes? Why do some people excel in sprint events whilst others make outstanding distance runners? Why is it so rare for one person to achieve the highest level in both?

Exercise poses some real challenges for the body. Many changes occur without our conscious thought, such as the finely controlled adjustments that ensure oxygen and fuel are supplied in sufficient quantities to the muscles. Fit sportspeople may not be aware of this unless working very hard, when the heart starts pounding and breathing becomes laboured. Overheating may be something they are much more aware of as they cover the kilometres. How does the body ensure that muscles are well supplied with oxygen and fuel, and how does it prevent the body temperature rising too high?

We are always being encouraged to take plenty of exercise. This is good advice with our increasingly sedentary lifestyles which are leading to an ever higher incidence of obesity, cardiovascular disease and related diseases. But can you overdo it? What are the consequences of overtraining or trying to improve on nature?

▲ **Figure 7.3** Our abilities have evolved over thousands of years as we adapted to our surroundings.

▲ **Figure 7.4** Paula Radcliffe winning gold for the marathon at the World Championships in 2005. She set a new world record of 2:15:25 in 2003.

Overview of the biological principles covered in this topic

At the start of the topic you will recall ideas from KS4 about joints and movement. This is then extended to provide a detailed understanding of the mechanism of contraction in skeletal muscle.

While studying respiration, you will revisit ideas from Topic 5 about energy transfer in biological systems, ATP, redox reactions and electron transport chains. The different energy systems are considered in detail. Building on your knowledge of the cardiovascular and ventilation systems from Topics 1 and 2 you will investigate the control of heart rate, lung volumes and breathing rate.

In this topic you will re-examine homeostasis and negative feedback from Topic 6 using the example of thermoregulation in a sporting context. You will return to the principles of the immune response introduced in Topic 6 in the context of immune suppression as a consequence of overtraining.

You will investigate the use of medical technology to enable participation of those with injuries or disabilities. You will discuss the moral and ethical issues surrounding the use of performance-enhancing substances in sport.

7.1 Getting moving

Before the starter's gun fires, the sprinter in Figure 7.5 is poised for action. He is crouched on the blocks, legs and ankles flexed. As the gun fires he pushes against the blocks to get maximum thrust and is up and running. Like the cheetah, with every step, muscles contract and relax to bend (**flex**) and straighten (**extend**) ankles, knees and hip joints, exerting a force against the ground to push his body forward.

Joints

▲ **Figure 7.5** An athlete prepared for action.

Joints and movement

Muscles bring about movement at a **joint**. Most movements are produced by the coordinated action of several muscles. Muscles shorten, pulling on the bone and so moving the joint. Muscles can only pull, they cannot push, so at least two muscles are needed to move a bone to and fro. A pair of muscles that work in this way are described as **antagonistic**. For example, when you flex your knee by contracting the hamstring muscles at the back of the thigh, the quadriceps at the front of the thigh relax and so are stretched. To extend the knee the quadriceps shorten while the hamstrings relax.

A muscle that contracts to cause extension of a joint is called an **extensor**; the corresponding **flexor** muscle contracts to reverse the movement.

Q7.1 In Figure 7.6 which muscle, A or B, brings about extension of the cat knee joint?

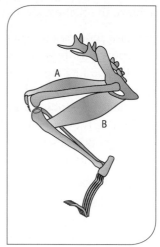

Joint structure

The hip, knee and ankle joints are examples of **synovial joints**; the bones that articulate (move) in the joint are separated by a cavity filled with **synovial fluid** which enables them to move freely. All synovial joints have the same basic structure as that shown in Figure 7.7. The bones are held in position by **ligaments** that control and restrict the amount of movement in the joint. **Tendons** attach muscles to the bones, enabling the muscles to power joint movement. **Cartilage** protects bones within joints. The function of each part of the joint is shown in Figure 7.7.

▲ **Figure 7.6** Muscles for flexion and extension of a cat knee joint.

Q7.2 **a** What properties of ligaments make them effective at holding the bones in place at a joint?
b What reduces the wear and tear due to friction in a mobile synovial joint?

Q7.3 Compare the simplified synovial joint in Figure 7.7 with the diagram of the human knee joint in Figure 7.8.
a Identify the key features of the knee joint labelled A to C.
b Which features of a synovial joint are not shown in this diagram?

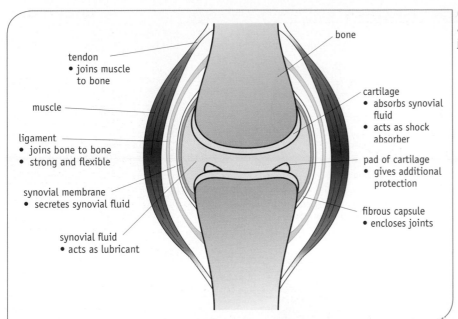

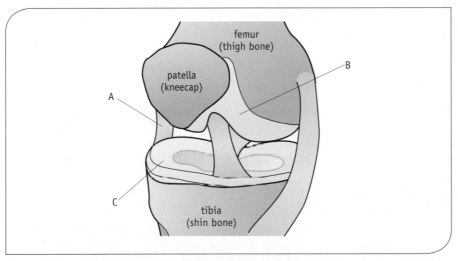

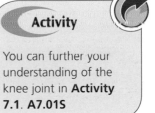

Figure 7.7 Where bones meet a joint is formed. A typical synovial joint allows bones to move freely.

Figure 7.8 The human knee joint.

Activity

You can further your understanding of the knee joint in **Activity 7.1. A7.01S**

Did you know? Other types of joint

Not all joints have the same structure. The plates of the skull are fixed together by fibrous tissue so very little movement occurs at these fixed joints. As a result the brain is protected inside a rigid bone casket. In the skull of a newborn baby the joints are not yet fixed; this allows the plates to move and the skull to deform (reversibly!) during birth. There are spaces between the skull bones, which fill in as the plates grow and fuse together.

A mother's pelvic bones are joined together by cartilage, so only slight movement is possible during childbirth; this is an example of a cartilaginous joint. There is also a wide variety of synovial joints allowing different degrees of extension, flexion and rotation, as shown in Figure 7.9.

(Continued)

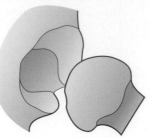

Ball-and-socket joint

A round head fits into a cup-shaped socket, e.g. the hip.

Gliding joint

Two flat surfaces slide over one another, e.g. the articular process between neighbouring vertebrae.

Hinge joint

A convex surface fits into a concave surface, e.g. the elbow.

Atlas bone supports the head.

Pivot joint

Part of one bone fits into a ring-shaped structure and allows rotation, e.g. the joint at the top of the spine.

There are also saddle joints and condyloid joints in which more complex concave and convex surfaces articulate.

◀ **Figure 7.9** Different types of synovial joint.

Did you know? What makes a cheetah so fast?

The cheetah's slender body, long legs and flexible spine allow it to run at great speed. The cheetah flexes its spine and rotates its shoulder blades forward and back which gives it a longer stride (Figure 7.10). Unlike other cats, the cheetah does not retract its claws into protective sheaths; the exposed claws, combined with hard ridges on their feet pads, function rather like an athlete's spikes. The disadvantage of having permanently exposed claws is that they become blunt, making them less good as running spikes and for catching prey. The latter is not too much of a problem for the cheetah which uses an enlarged claw on the inner sides of the front legs to bring down moving prey.

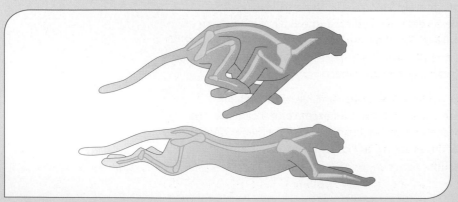

◀ **Figure 7.10** The cheetah flexes its spine.

Muscles

How do muscles work?

The key to how muscles function is in their internal structure. Muscles are made up of bundles of **muscle fibres**. Each fibre is a muscle cell (Figures 7.11 and 7.12) and can be several centimetres in length.

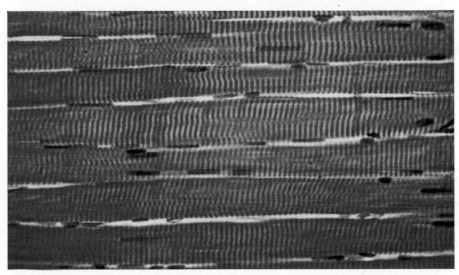

🔺 **Figure 7.11** Human muscle fibres. Each muscle fibre is a single cell that is wider than this photograph.

Apart from the considerable length of the cells, what do you notice about the muscle fibres that is unusual? You should spot that each cell has several nuclei (is **multinucleate**). This occurs because a single nucleus could not effectively control the metabolism of such a long cell. During prenatal development, several cells fuse together forming an elongated muscle fibre. The muscle cells are also striped; as we shall see, this is an important feature related to their ability to contract.

Q7.4 Why would a single nucleus be unable to control the metabolism of a long thin cell like a muscle fibre?

Inside a muscle fibre

Within each muscle fibre are numerous **myofibrils** (Figure 7.12). These are made up of a series of contractile units called **sarcomeres**, as shown in Figures 7.13 and 7.14.

A sarcomere is made up two types of protein molecule: thin filaments made up mainly of the protein **actin** and thicker ones made of the protein **myosin**. Contractions are brought about by coordinated sliding of these protein filaments within the muscle cell sarcomeres.

The arrangement of the filaments in a sarcomere is shown in Figures 7.13 and 7.14. The proteins overlap and give the muscle fibre its characteristic striped (striated) appearance under the microscope (Figure 7.11). Where actin filaments occur on their own, there is a *light* band on the sarcomere. Where both actin and myosin filaments occur there is a *dark* band. Where only myosin filaments occur there is an intermediate-coloured band (Figure 7.14B).

Weblink

Compare muscle cells with the typical animal cell in the interactive tutorial on cell structure and function in **Activity 3.1**.

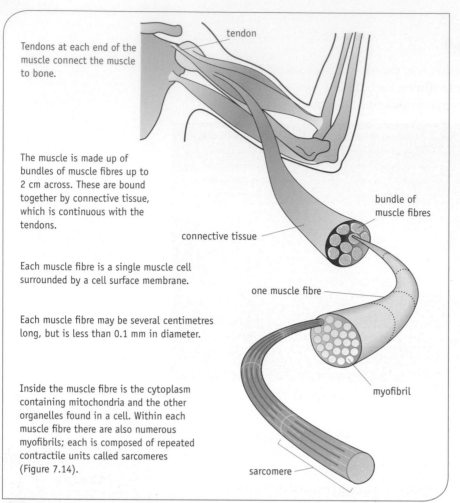

Tendons at each end of the muscle connect the muscle to bone.

tendon

bundle of muscle fibres

connective tissue

The muscle is made up of bundles of muscle fibres up to 2 cm across. These are bound together by connective tissue, which is continuous with the tendons.

Each muscle fibre is a single muscle cell surrounded by a cell surface membrane.

one muscle fibre

Each muscle fibre may be several centimetres long, but is less than 0.1 mm in diameter.

myofibril

Inside the muscle fibre is the cytoplasm containing mitochondria and the other organelles found in a cell. Within each muscle fibre there are also numerous myofibrils; each is composed of repeated contractile units called sarcomeres (Figure 7.14).

sarcomere

▲ **Figure 7.12** The arrangement of muscle fibres within muscles.

▲ **Figure 7.13** Electron micrograph showing the banding pattern of sarcomeres.

When the muscle contracts, the actin moves between the myosin as shown in Figure 7.14C, shortening the length of the sarcomere and hence the length of the muscle.

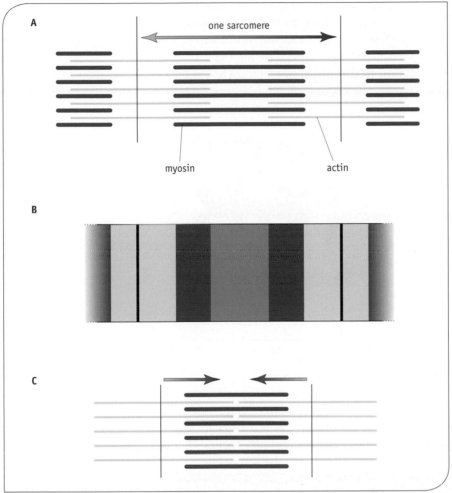

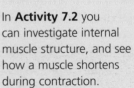

Activity

In **Activity 7.2** you can investigate internal muscle structure, and see how a muscle shortens during contraction.
A7.02S

▲ **Figure 7.14 A** The arrangement of actin and myosin filaments within a single muscle sarcomere when relaxed (extended). **B** The banding patterns created on an extended muscle myofibril. **C** The arrangement when the muscle contracts.

Q7.5 How many myofibrils are shown in Figure 7.13?

Q7.6 Look at the banding pattern on the contracted muscle fibre shown in Figure 7.15. Explain what has happened to the central, intermediate-coloured band visible on an extended muscle in Figure 7.14B.

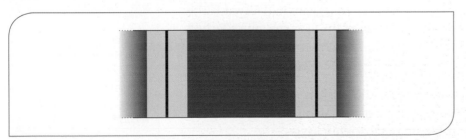

▲ **Figure 7.15** The banding pattern of a sarcomere when contracted.

How the sarcomere shortens

Actin molecules are associated with two other protein molecules called **troponin** and **tropomyosin**. Myosin molecules are shaped rather like golf clubs; the club shafts lie together as a bundle, with the heads protruding along their length (Figure 7.16). In a contraction, when the muscle shortens, the change in orientation of the myosin heads brings about the movement of actin. The myosin heads attach to the actin and dip forward, sliding the actin over the myosin. This is called the **sliding filament theory** of muscle action; the detailed sequence of events is described below and shown in Figure 7.17.

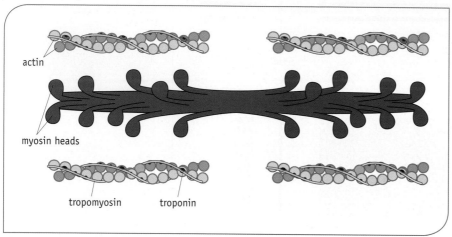

▲ **Figure 7.16** The myosin heads project on either side of the myosin molecule.

The sliding filament theory

When a nerve impulse arrives at a **neuromuscular junction**, calcium ions (Ca^{2+}) are released from the **sarcoplasmic reticulum**. This is a specialised type of endoplasmic reticulum: a system of membrane-bound sacs around the myofibrils (Figure 7.17A). The calcium ions diffuse through the sarcoplasm (the name given to cytoplasm in a muscle cell). This initiates the movement of the protein filaments as follows (Figure 7.17B):

- Ca^{2+} attaches to the troponin molecule, causing it to move.
- As a result, the tropomyosin on the actin filament shifts its position, exposing myosin binding sites on the actin filaments.
- Myosin heads bind with myosin binding sites on the actin filament, forming cross-bridges.
- When the myosin head binds to the actin, ADP and P_i on the myosin head are released.
- The myosin changes shape, causing the myosin head to nod forward. This movement results in the relative movement of the filaments; the attached actin moves over the myosin.
- An ATP molecule binds to the myosin head. This causes the myosin head to detach.
- An ATPase on the myosin head hydrolyses the ATP, forming ADP and P_i.
- This hydrolysis causes a change in shape of the myosin head. It returns to its upright position. This enables the cycle to start again.

Activity

Building a model in **Activity 7.3** will let you check your understanding of the sliding filament theory. **A7.03S**

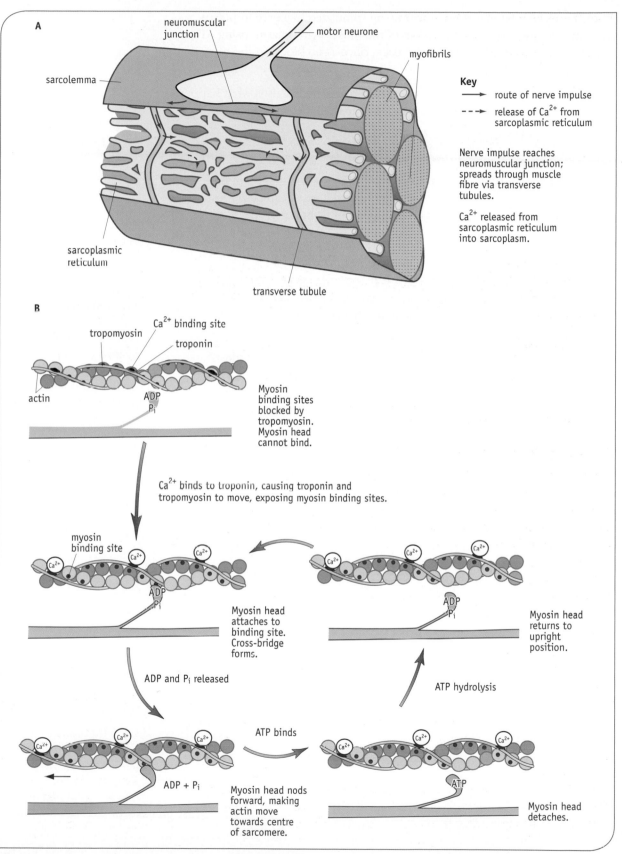

A

neuromuscular junction

motor neurone

myofibrils

sarcolemma

Key

→ route of nerve impulse

- - → release of Ca^{2+} from sarcoplasmic reticulum

Nerve impulse reaches neuromuscular junction; spreads through muscle fibre via transverse tubules.

Ca^{2+} released from sarcoplasmic reticulum into sarcoplasm.

sarcoplasmic reticulum

transverse tubule

B

tropomyosin

Ca^{2+} binding site

troponin

actin

ADP P$_i$

Myosin binding sites blocked by tropomyosin. Myosin head cannot bind.

Ca^{2+} binds to troponin, causing troponin and tropomyosin to move, exposing myosin binding sites.

myosin binding site

Ca^{2+}

ADP P$_i$

Myosin head attaches to binding site. Cross-bridge forms.

ADP and P$_i$ released

Ca^{2+}

ADP P$_i$

Myosin head returns to upright position.

ATP hydrolysis

Ca^{2+}

ADP + P$_i$

Myosin head nods forward, making actin move towards centre of sarcomere.

ATP binds

Ca^{2+}

ATP

Myosin head detaches.

Figure 7.17 The sliding filament theory of muscle contraction.

When a muscle relaxes it is no longer being stimulated by nerve impulses. Calcium ions are actively pumped out of the muscle sarcoplasm, using ATP. The troponin and tropomyosin move back, once again blocking the myosin binding site on the actin.

In the absence of ATP the cross-bridges remain attached. This is what happens in rigor mortis; any contracted muscles become rigid (Topic 6, page 80).

Did you know? Other types of muscle

Muscles found in such places as the gut wall, blood vessels and iris of the eye are known as smooth muscle: their fibres do not appear striped. These small cells with a single nucleus have a similar mechanism of contraction using myosin and actin protein filaments, but they are not arranged in the same way as occurs in skeletal muscle. Smooth muscle fibre contractions are slower and longer lasting, and the fibres fatigue very slowly if at all. Gap junctions – intercellular channels less than 2 nm diameter – between the smooth muscle cells give cytoplasmic continuity between the cells. This allows chemical and electrical signals to pass between adjacent cells and allows synchronised contraction.

The heart walls are made of specialised muscle fibres called cardiac muscle; these are striped and interconnected by gap junctions to ensure that a coordinated wave of contraction occurs in the heart. Cardiac muscle fibres do not fatigue. Neither smooth nor cardiac muscle are under conscious control – unless you become extremely good at certain forms of yoga.

Checkpoint

7.1 Produce a flowchart of the steps that must occur to bring about contraction of a sarcomere.

Extension

Find out what the method of wrinkle control (BOTOX™) and a lethal form of food poisoning (botulism) have in common in **Extension 7.1**. **X7.01S**

7.2 Energy for action

Just staying alive, even if you're not doing anything active, uses a considerable amount of energy. This minimum energy requirement is called the **basal metabolic rate** (**BMR**) (see Topic 1). BMR is a measure of the minimum energy requirement of the body at rest to fuel basic metabolic processes, and is measured in kJ g^{-1} h^{-1}.

BMR is measured by recording oxygen consumption under strict conditions; no food is consumed for 12 hours before measurement with the body totally at rest in a thermostatically controlled room. BMR is roughly proportional to the body's surface area. It also varies between individuals depending on their age and gender. Percentage body fat seems to be important in accounting for these differences.

Q7.7 Women generally have a higher percentage of body fat than men. Why might this account for the fact that women generally have a lower BMR than men?

Physical activity increases the body's total daily energy expenditure. Energy is needed for muscle contraction to move the body around, as you saw in Activity 1.18. An elite marathon runner uses energy at a rate of about 1.75 kJ s^{-1}, whereas a sprinter expends around 4 kJ s^{-1}. How does the muscle cell deal with these very different energy demands?

Releasing energy

Food is the source of energy for all animal activity. The main energy sources for most people are carbohydrates and fats, which have either just been absorbed from the gut or have been stored around the body. A series of enzyme-controlled reactions, known as **respiration**, is linked to ATP synthesis. As you saw in Topic 5, cells use the molecule ATP as an energy carrier molecule. This is the cell's energy currency, coupling energy-yielding reactions with energy-requiring reactions, as in the muscle contraction described earlier in this topic.

◀ **Figure 7.18** ATP is a universal molecule used by living organisms. Here a firefly releases energy from ATP to create luminescence in its attempts to attract mates. Magnification ×3.

ATP is created from ADP by the addition of inorganic phosphate (P_i). In order to make ATP, phosphate must be torn away from water molecules, and this reaction requires energy. ATP in water is higher in energy than ADP and phosphate ions in water, so ATP in water is a way of storing chemical potential energy. ATP separates the phosphate from the water, but they can be brought together in an energy-yielding reaction each time energy is needed for reactions within the cell.

When one phosphate group is removed from ATP, adenosine diphosphate forms. A small amount of energy is required to break the bond holding the end phosphate in the ATP. Once removed, the phosphate group becomes hydrated. A lot of energy is released as bonds form between water and phosphate. This energy can be used to supply energy-requiring reactions in the cell. Some of the energy transferred during hydration of phosphate from ATP will raise the temperature of the cell; some is available to drive other metabolic reactions such as muscle contraction, protein synthesis or active transport. The hydrolysis of ATP is coupled to these other reactions:

ATP in water → ADP in water + hydrated P_i + energy transferred

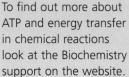

Support

To find out more about ATP and energy transfer in chemical reactions look at the Biochemistry support on the website.

Carbohydrate oxidation

If exercise is low intensity, for example in long-distance walking and running, enough oxygen is supplied to cells to enable ATP to be regenerated through aerobic respiration of fuels. Fats and carbohydrates, like glucose, are oxidised to carbon dioxide and water; you are probably familiar with the summary equation for aerobic respiration:

$$C_6H_{12}O_6 + 6O_2 \rightarrow 6CO_2 + 6H_2O + \text{energy released}$$

In Topic 5, photosynthesis was described as a process which separates hydrogen from oxygen by pulling apart water molecules. The hydrogen from water is stored by combining it with carbon dioxide to form carbohydrate.

In aerobic respiration, the hydrogen stored in glucose is brought together with oxygen to form water again. The bonds between hydrogen and carbon in glucose are not as strong as the bonds between hydrogen and oxygen in water. So the input of energy to break the bonds in glucose and oxygen is not as great as the energy released when the bonds in carbon dioxide and water are formed. Overall, there is a release of energy, and this can be used to generate ATP.

Glucose and oxygen are not brought together directly, as this would release large amounts of energy which may cause damage to the cell. Glucose is pulled apart in a series of small steps. Carbon dioxide is pulled off and released as a waste product. Hydrogen is pulled off the glucose and is eventually reunited with oxygen to release large amounts of energy as water is formed.

Having understood the significance of the overall reaction and energy changes in respiration, you need to look in more detail at the mechanisms that take place during this process.

Glycolysis first

The initial stages of carbohydrate breakdown, known as **glycolysis**, occur in the cytoplasm of cells, including the sarcoplasm of muscle cells.

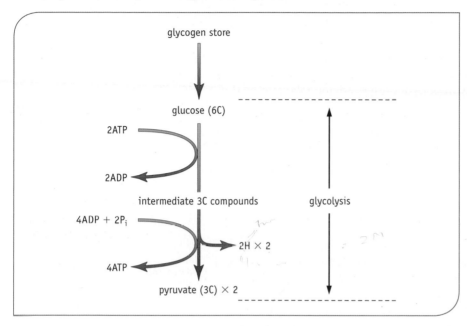

Figure 7.19 The glycolysis reactions in respiration are anaerobic.

Stores of glycogen (a polymer of glucose, see Topic 1) in muscle or liver cells are converted to glucose. Glucose is a good fuel – it can potentially yield $2880 \, kJ \, mol^{-1}$ – but it is quite stable and unreactive. Therefore, the first reactions of glycolysis need an *input* of energy from ATP to get things started. Two phosphate groups are added to the glucose from two ATP molecules, and this increases the reactivity of the glucose. It can now be split into two molecules of 3-carbon (3C) compounds. It is rather like lighting a candle; a match provides an initial input of energy before energy can be released from the fuel (candle wax).

Each intermediate 3C sugar is oxidised, producing the 3-carbon compound **pyruvate**; two hydrogen atoms are removed during the reaction and taken up by the **coenzyme NAD** (**nicotinamide adenine dinucleotide**), a non-protein organic molecule. The fate of these hydrogens and their role in ATP synthesis is explained on pages 145–6.

Glucose is at a higher energy level than the pyruvate so, on conversion, some energy becomes available for the direct creation of ATP. Phosphate from the intermediate compounds is transferred to ADP, creating ATP. This is called **substrate-level phosphorylation**, because energy for the formation of ATP comes from the substrates; in this case the intermediate compounds are the substrates (Figure 7.19). Four ATP molecules are created but because 2ATPs are used to phosphorylate the glucose the net gain is just 2ATPs.

In summary, glycolysis reactions yield a net gain of 2ATPs, 2 pairs of hydrogen atoms, and two molecules of 3-carbon pyruvate, as shown in Figure 7.19.

Q7.8 Why are the glycolysis reactions described as anaerobic?

What happens to the pyruvate depends on the availability of oxygen.

The fate of pyruvate if oxygen is available

If oxygen is available, the 3C pyruvate created at the end of glycolysis passes into the mitochondria. There it is completely oxidised forming carbon dioxide and water (see Figure 7.20).

The link reaction

In the first step, known as the **link reaction**, pyruvate is:

- **decarboxylated** (carbon dioxide is released as a waste product)
- **dehydrogenated** (two hydrogens are removed and taken up by the coenzyme NAD).

The resulting 2-carbon molecule combines with coenzyme A to form **acetyl coenzyme A** (**acetyl CoA** for short). As we shall see, the two hydrogen atoms released are involved in ATP formation. The coenzyme A carries the 2C acetyl groups to the **Krebs cycle**.

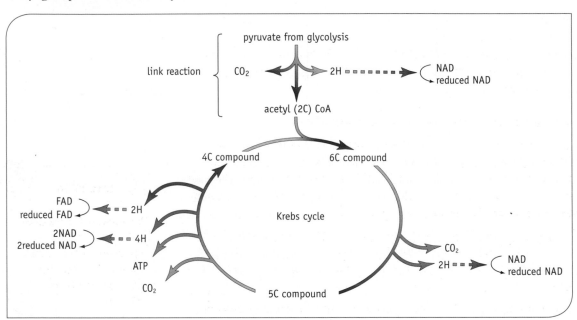

🔺 **Figure 7.20** The reactions involved in the breakdown of glucose in aerobic respiration. Each glucose provides two pyruvates so the cycle turns twice per glucose molecule. The hydrogens are taken up by hydrogen acceptors, the coenzymes NAD and FAD which become reduced NAD and reduced FAD.

Krebs cycle

The 2-carbon compound from each acetyl CoA combines with a 4-carbon compound to create one with 6 carbons. In a circular pathway of reactions, the original 4-carbon compound is recreated. In these reactions, two steps involve decarboxylation with the formation of carbon dioxide. Four steps involve dehydrogenation, the removal of pairs of hydrogen atoms. In addition, one of the steps in the cycle also involves substrate-level phosphorylation with direct synthesis of a single ATP (see glycolysis substrate-level phosphorylation on page 143). This circular pathway of reactions (Figure 7.20) is known as the Krebs cycle, named after Sir Hans Krebs who worked out the cycle of reactions. The Krebs cycle takes place in the mitochondrial matrix (Figure 7.22) where the enzymes that catalyse the reactions are located.

Activity

In **Activity 7.4** you can produce a summary diagram of glycolysis and the Krebs cycle.
A7.04S

The chemical reactions inside cells are controlled by enzymes. There are four important types of reaction in the Krebs cycle:

- phosphorylation reactions which add phosphate, e.g. $ADP + P_i \rightarrow ATP$
- decarboxylation reactions which break off carbon dioxide, e.g. pyruvate $\rightarrow$ acetyl CoA + CO_2
- redox reactions where substrates are oxidised and reduced, e.g.
  (i) oxidised NAD + electrons $\rightarrow$ reduced NAD
  (ii) pyruvate $\rightarrow$ acetyl CoA + 2H (dehydrogenation).

When a molecule is oxidised, one or more electrons, e^-, are lost from this substrate molecule, and the molecule that accepts these electrons is reduced. See Topic 5, page 32. When a molecule loses hydrogen (dehydrogenation) it is also oxidised, and the molecule that gains the hydrogen is reduced. In the example above, pyruvate loses hydrogen to form oxidised acetyl CoA. Adding and removing oxygen is another type of redox reaction.

In summary, each 2-carbon molecule entering the Krebs cycle results in the production of two carbon dioxide molecules, one molecule of ATP by substrate-level phosphorylation, and eight pairs of hydrogen atoms. These hydrogen atoms are subsequently involved in ATP production via the **electron transport chain** described below.

Fate of the hydrogens – the electron transport chain

Hydrogen atoms released during glycolysis, the link reaction and the Krebs cycle are taken up by coenzymes as summarised in Figure 7.22. For most hydrogens produced, the coenzyme NAD is the hydrogen acceptor, although those released at one stage in the Krebs cycle are accepted by the coenzyme **FAD (flavine adenine dinucleotide)** rather than NAD.

When a coenzyme accepts a hydrogen with its electron the coenzyme is reduced, becoming reduced NAD or reduced FAD. The reduced coenzyme 'shuttles' the hydrogen atoms to the electron transport chain on the mitochondrial inner membrane. As illustrated in Figure 7.21, each hydrogen atom's electron and proton (H^+) then separate, with the electron passing along a chain of electron carriers in the inner mitochondrial membrane. This is known as the electron transport chain.

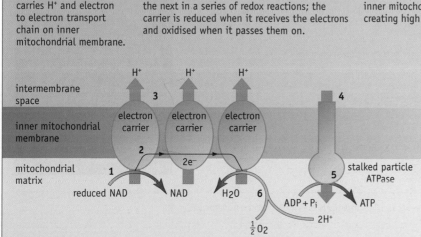

1 Reduced coenzyme carries H^+ and electron to electron transport chain on inner mitochondrial membrane.

2 Electrons pass from one electron carrier to the next in a series of redox reactions; the carrier is reduced when it receives the electrons and oxidised when it passes them on.

3 Protons (H^+) move across the inner mitochondrial membrane creating high H^+ concentrations.

4 H^+ diffuse back into the mitochondrial matrix down the electrochemical gradient.

5 H^+ diffusion allows ATPase to catalyse ATP synthesis.

6 Electrons and H^+ ions combine to form hydrogen atoms which then combine with oxygen to create water. If the supply of oxygen stops, the electron transport chain and ATP synthesis also stop.

Figure 7.21 The electron transport chain and chemiosmosis result in ATP synthesis.

ATP synthesis by chemiosmosis

How does the electron transport chain lead to ATP synthesis? This is explained by the **chemiosmotic theory** summarised in Figure 7.21. When it was first proposed, the chemiosmotic theory was ignored by the scientific establishment. Nowadays, it is very widely accepted. Using energy released as electrons pass along the electron transport chain, protons (hydrogen ions) originating from the hydrogen atoms released in glycolysis or the Krebs cycle are moved from the matrix across the inner mitochondrial membrane into the intermembrane space. This creates a steep **electrochemical gradient** across the inner membrane. There is a large difference in concentration of H^+ across the membrane, and a large electrical difference; the intermembrane space is more positive than the matrix.

The hydrogen ions diffuse down this electrochemical gradient through hollow protein channels in stalked particles on the membrane. As the hydrogen ions pass through the channel, ATP synthesis is catalysed by ATPase located in each stalked particle. The hydrogen ions cause a conformational change (change in shape) in the enzyme's active site so the ADP can bind.

Within the matrix, the H^+ and electrons recombine to form hydrogen atoms. These combine with oxygen to form water. The oxygen, acting as the final carrier in the electron transport chain, is thus reduced. This method of synthesising ATP is known as **oxidative phosphorylation**.

Q7.9 Why is the synthesis of ATP via the electron transport chain termed 'oxidative phosphorylation'?

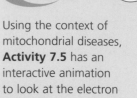

Activity

Using the context of mitochondrial diseases, **Activity 7.5** has an interactive animation to look at the electron transport chain and chemiosmosis. **A7.05S**

Support

To find out more about coenzymes, redox reactions and electron transport chains visit the Biochemistry support on the website.

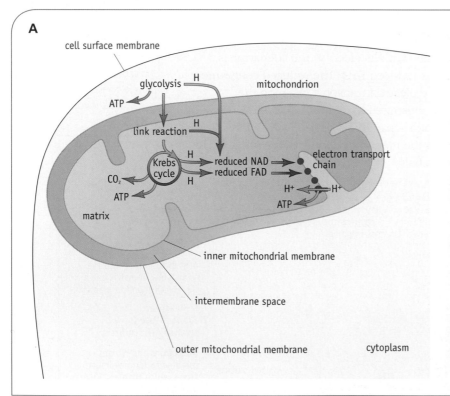

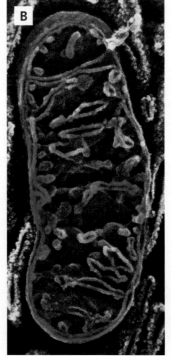

🔺 **Figure 7.22 A** A summary of where the respiration reactions occur. **B** Coloured electron micrograph of a mitochondrion.

How much ATP is produced?

Each reduced NAD which is reoxidised results in the production of 3ATP. Each reduced FAD transfers sufficient energy to produce 2ATP. The total number of ATP produced in the different stages of the cycle can be calculated using Figures 7.19 and 7.20. Try Question 7.10 to check whether you can work it out.

Q7.10   Look at Figures 7.19 and 7.20 and work out the total number of ATPs produced by: **a** substrate-level phosphorylation and **b** oxidative phosphorylation when one glucose molecule passes through the respiration reactions.

Combining the values determined in Question 7.10, you should find that 38ATP are produced from the complete aerobic respiration of each glucose molecule. Complete oxidation (e.g. by combustion) of one mole of glucose molecules releases 2880 kJ. The 38 moles of ATP molecules formed in aerobic respiration of one mole of glucose molecules can release only 1163 kJ, just 40% of the total potential chemical energy stored in the glucose. The remaining energy raises the temperature of the cell. This helps to increase the rate of metabolic reactions and, in mammals and birds, to maintain core body temperature.

Fatty acid oxidation

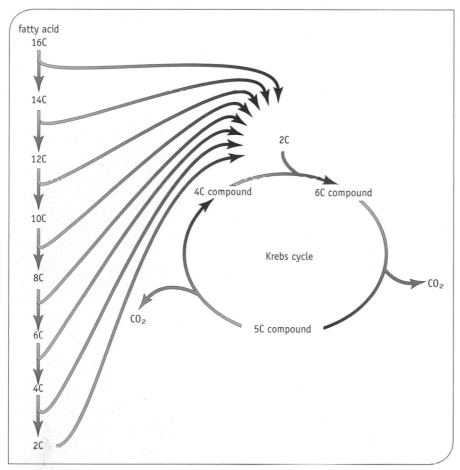

Figure 7.23 Oxidation of fatty acids occurs via the Krebs cycle.

Fats can also be respired to release energy, and are a richer energy store than carbohydrates – glycogen and glucose release $17\,kJ\,g^{-1}$ whereas triglycerides release $37\,kJ\,g^{-1}$. In fatty acid oxidation, the glycerol and fatty acids which make up triglycerides are separated. The fatty acids are broken down in a series of reactions, each generating the same 2-carbon compound, and these can be fed into the Krebs cycle for oxidation (Figure 7.23).

Because fatty acids can only be respired through the Krebs cycle, fats can only be a fuel for aerobic respiration, and cannot be used when oxygen is not available. Glucose can be respired aerobically or anaerobically.

Did you know? Using proteins

Some modern athletes believe, like their ancient Olympian counterparts, that a high-protein diet is the key to successful competition. However, excess intake of any fuel, including protein, is converted into fat. Amino acids will be built into muscle protein only if there is training overload. The body may use the excess protein as fuel through respiration but to do this the amino acids must first be deaminated, producing urea. Water is needed to excrete the extra urea. As urine output increases, the body's fluid requirements also increase.

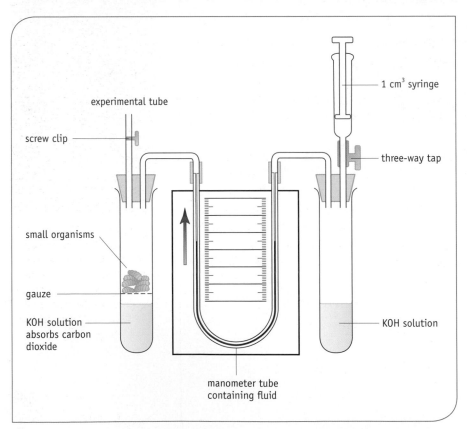

🔺 **Figure 7.24** As the living organisms in the experimental tube take up oxygen the fluid in the manometer will move in the direction of the arrow. The potassium hydroxide (KOH) solution absorbs any carbon dioxide produced by the organisms through respiration. The second tube compensates for any changes in volumes due to variations in gas pressure or temperature inside the apparatus.

Rate of respiration

In small organisms the rate of aerobic respiration can conveniently be determined by measuring the uptake of oxygen using a respirometer (Figure 7.24). Respiration is a series of enzyme-catalysed reactions, so its rate will be influenced by any factor that affects the rate of these enzyme-controlled reactions. For example, enzyme concentration, substrate concentration, temperature and pH will all affect the rate of respiration.

The concentration of ATP in the cell also has a role in the control of respiration. ATP inhibits the enzyme in the first step of glycolysis, the phosphorylation of glucose. The enzyme responsible for glucose phosphorylation can exist in two different forms. In the presence of ATP the enzyme has a shape that makes it inactive; it cannot catalyse the reaction. As ATP is broken down the enzyme is converted back to the active form and catalyses the phosphorylation of glucose again. This is known as end-point inhibition: the end product inhibits an early step in the metabolic pathway, so controlling the whole process.

Q7.11 What is the advantage of having a system of enzyme-controlled reactions to transfer energy from food fuels?

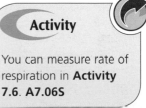

Activity

You can measure rate of respiration in **Activity 7.6. A7.06S**

The fate of pyruvate without oxygen

Anaerobic respiration

At the start of any exercise and during intense exercise, for example a 400 m race, oxygen demands in the cells exceed supply (Figure 7.25). Without oxygen to accept the hydrogen ions and electrons, the electron transport chain ceases and the reduced NAD created during glycolysis, the link reaction and the Krebs cycle is not oxidised. Without a supply of oxidised NAD the respiration reactions cannot continue.

However, in the absence of oxygen it is possible to oxidise the reduced NAD created during glycolysis. The pyruvate produced at the end of glycolysis is reduced to **lactate** and the oxidised form of NAD is regenerated. In this way anaerobic respiration allows the athlete to continue by partially breaking down glucose to make a small amount of ATP (Figure 7.26). The net yield is just two ATP molecules per glucose molecule – in other words, only 61 kJ of energy made available in the form of ATP. This process has only 2% efficiency.

▲ **Figure 7.25** The women's world record for running 400 m is 47.6 seconds. The intensity of the activity is such that the muscle cells do not get enough oxygen for aerobic respiration. How do such athletes fuel their performance?

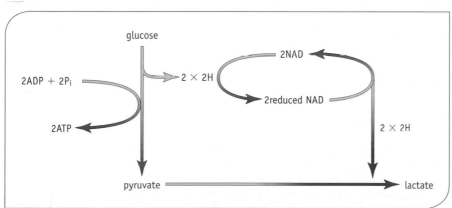

▲ **Figure 7.26** Anaerobic respiration.

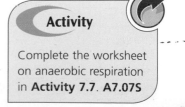

Activity

Complete the worksheet on anaerobic respiration in **Activity 7.7. A7.07S**

The end-product of anaerobic respiration is lactate, which builds up in the muscles and must be disposed of later. Animal tissues can tolerate quite high levels of lactate, but lactate forms lactic acid in solution. This means that as lactate accumulates, the pH of the cell falls, inhibiting the enzymes which catalyse the glycolysis reactions. The glycolysis reactions and the physical activity that depends on them cannot continue.

The effect of lactate build-up

Enzymes function most efficiently over a narrow pH range. Many of the amino acids that make up an enzyme have positively or negatively charged groups. As hydrogen ions from the lactic acid accumulate in the cytoplasm, they neutralise the negatively charged groups in the active site of the enzyme. The attraction between charged groups on the substrate and in the active site will be affected (Figure 7.27). The substrate may no longer bind to the enzyme's active site.

substrate

active site

As pH falls, hydrogen ions prevent the substrate from being attached to the charged groups in the active site.

🔺 **Figure 7.27** As lactate builds up within cells, the pH eventually falls, inhibiting the glycolysis reactions. The athlete, person sprinting for a bus or any other exhausted animal has to slow down or stop so that oxygen supply can meet demand. Lactate build-up may lead to muscular cramp.

Getting rid of lactate

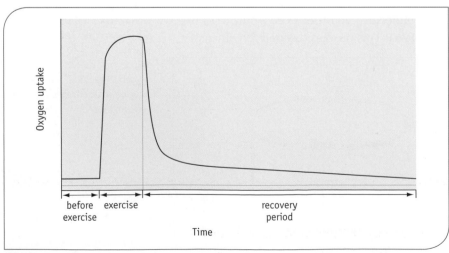

🔺 **Figure 7.28** Oxygen uptake in the recovery period.

After a period of anaerobic respiration, most of the lactate is converted back into pyruvate and is oxidised directly to carbon dioxide and water via the Krebs cycle, thus releasing energy to synthesise ATP. As a result, oxygen uptake is greater than normal in the recovery period after exercise (see Figure 7.28). This excess oxygen requirement is called the **oxygen debt**, or **post-exercise oxygen consumption**. It is needed to fuel the oxidation of lactate. Some lactate may also be converted into glycogen and stored in the muscle or liver.

Checkpoint

7.2 Draw up a table of comparisons between aerobic and anaerobic respiration.

Q7.12 Why are athletes in training advised not to lie down after strenuous exercise, but rather to aim for active recovery through gentle exercise?

Did you know? Alcoholic fermentation

Yeast cells adopt a different tactic from animals in anaerobic conditions. They reduce pyruvate to ethanol and CO_2, using the hydrogen from reduced NAD, thus recreating oxidised NAD and allowing glycolysis to continue (Figure 7.29). This anaerobic respiration, also known as fermentation, is exploited in the brewing industry. Yeast cells are facultative anaerobes. Aerobes are organisms that respire using oxygen; anaerobes respire without using oxygen. Facultative anaerobes use either aerobic or anaerobic respiration depending on the conditions.

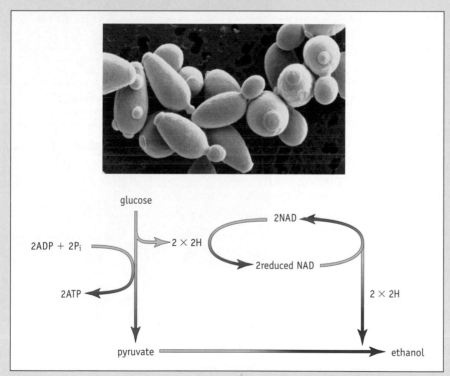

Figure 7.29 Yeast cells (top) can respire both aerobically and anaerobically. In the absence of oxygen they use alcohol fermentation. Photo magnification ×1050.

Supplying instant energy

Cells store only a tiny amount of ATP, enough in humans to allow a couple of seconds of explosive, all-out exercise. The recharging of the ATP store therefore has to be very rapid.

At the start of any type of exercise the immediate regeneration of the ATP is achieved using **creatine phosphate** (sometimes also called phosphocreatine, PC). This is a substance stored in muscles which can be hydrolysed to release energy. This energy can be used to regenerate ATP from ADP and phosphate, the phosphate coming from the creatine phosphate itself. Creatine phosphate breakdown begins as soon as exercise starts (triggered by the formation of ADP). The reactions can be represented as:

$$\text{creatine phosphate} \rightarrow \text{creatine} + P_i$$
$$ADP + P_i \rightarrow ATP$$

These can be summarised as:

$$\text{creatine phosphate} + ADP \rightarrow \text{creatine} + ATP$$

The reactions do not require oxygen, and provide energy for about 6–10 seconds of intense exercise. This is known as the ATP/PC system and is relied upon for regeneration of ATP during bursts of intense activity, for example when throwing or sprinting (Figure 7.30).

Later, creatine phosphate stores can be regenerated from ATP when the body is at rest.

Three energy systems

At the start of any exercise, aerobic respiration cannot meet the demands for energy because the supply of oxygen to the muscles is insufficient (Figure 7.31). The lungs and circulation are not delivering oxygen quickly enough, and ATP will be regenerated without using oxygen. First the ATP/PC system and then the anaerobic respiration system allow ATP regeneration.

In endurance-type exercise, a rapidly increased blood supply to the muscles ensures higher oxygen supply to the muscle cells. Aerobic respiration can regenerate ATP as quickly as it is broken down. This allows the exercise to be sustained for long periods. The relative contributions of the three different systems to the regeneration of ATP during exercise are shown in Figure 7.32.

Q7.13 Will the trained or untrained athlete create a larger oxygen deficit during the exercise period shown on the graph in Figure 7.31? Give a reason for your answer.

Q7.14 Table 7.1 shows the percentage contributions of the three energy systems during various sports. Which represents:

a volleyball **b** hockey **c** long-distance running?

▲ **Figure 7.30** These sprinters rely almost entirely on the ATP/PC system to supply the energy burst for their 10 second race.

Sport	ATP/PC	Anaerobic glycolysis to lactate	Aerobic respiration
A	60	20	20
B	90	10	0
C	10	20	70

◀ **Table 7.1** Percentage contributions of the three energy systems during various sports.

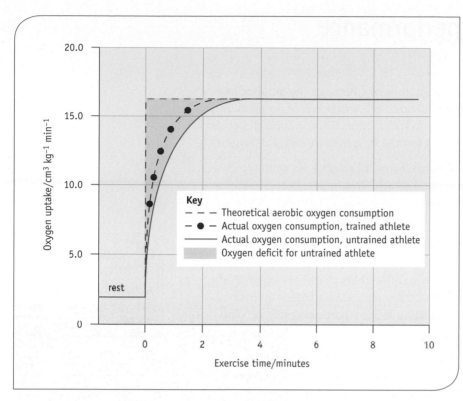

The oxygen deficit is the difference between the actual oxygen consumption and the theoretical oxygen consumption had the exercise been completed entirely aerobically.

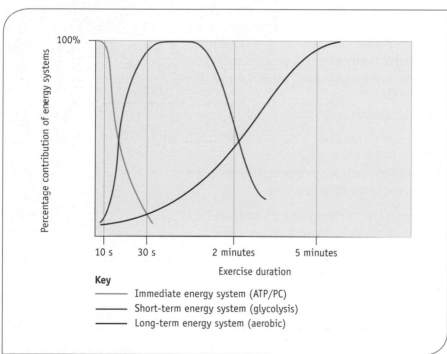

Figure 7.32 The relative contribution of anaerobic and aerobic respiration during exercise.

Q7.15 State which energy system:
a a cheetah will use in its sprint to catch a prey
b a wildebeest will use during the majority of its migration.

Q7.16 A cheetah is a carnivore, so its diet is largely protein and fat. How might it gain carbohydrate stores for use in anaerobic respiration?

7.3 Peak performance

The ability to undertake prolonged periods of strenuous but submaximal activity (e.g. running but not flat-out sprinting) is dependent on maintaining a continuous supply of ATP for muscle contraction. This, in turn, depends on **aerobic capacity**, that is the ability to take in, transport and use oxygen.

At rest we consume about 0.2–0.3 dm^3 (litres) of oxygen per minute. This is known as $\dot{V}O_2$ ($\dot{V}$ means volume per minute). This increases to 3–6 dm^3 a minute during maximal aerobic exercise, known as $\mathbf{\dot{V}O_2(max)}$. $\dot{V}O_2(max)$ is often expressed in units of ml min^{-1} kg^{-1} of body mass. Successful endurance athletes have a higher $\dot{V}O_2(max)$.

$\dot{V}O_2(max)$ is dependent on the efficiency of uptake and delivery of oxygen by the lungs and cardiovascular system, and the efficient use of oxygen in the muscle fibres. A fit person can work for longer and at a higher intensity using aerobic respiration without accumulating lactate than can someone who does not undertake regular aerobic exercise.

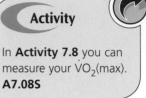

Activity

In **Activity 7.8** you can measure your $\dot{V}O_2(max)$. **A7.08S**

Most of us have run at some time and ended up breathing heavily with a pounding heart, or have seen a pet dog collapse panting at the end of a run. Without any conscious thought, the cardiovascular and ventilation systems adjust to meet the demands of the exercise, ensuring that enough oxygen and fuel reach the muscles, and removing excess carbon dioxide and lactate. These systems are also important in redistribution of energy in temperature control. The major changes are to **cardiac output** (see below), breathing rate and the depth of breathing. When running, adequate oxygen supply is maintained by:

- increasing cardiac output
- faster rate of breathing
- deeper breathing.

The more efficient the cardiovascular and ventilation systems, the better suited an individual will be to aerobic endurance-type exercise.

Look back at Topic 1 to refresh your memory of the structure and function of the heart, and Topic 2 to remind yourself about lungs and gas exchange.

Cardiac output

The volume of blood pumped by the heart in one minute is called the cardiac output. This increases during exercise. At rest this is approximately 5 dm^3 per minute in both trained and untrained individuals, but it can rise to about 30 dm^3 min^{-1} in a trained athlete making maximum effort.

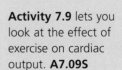

Activity

Activity 7.9 lets you look at the effect of exercise on cardiac output. **A7.09S**

The cardiac output depends on the volume of blood ejected from the left ventricle (the **stroke volume**) and the **heart rate**:

cardiac output = stroke volume (SV) × heart rate (HR)

Stroke volume

The stroke volume is the volume of blood pumped out of the left ventricle each time the ventricle contracts, measured in cm^3. (The volume pumped from both the left and right ventricles is virtually identical. Think about it!) For most adults at rest, about 50–90 cm^3 is pumped into each of the pulmonary artery and aorta when the ventricles contract.

The heart draws blood into the atria as it relaxes during diastole. How much blood the heart pumps out with each contraction is determined by how much blood is filling the heart, that is, the volume of blood returning to the heart from the body. During exercise there is greater muscle action so more blood returns to the heart in what is known as the **venous return**. In diastole during exercise the heart fills with a larger volume of blood. The heart muscle is stretched to a greater extent, causing it to contract with a greater force, and so more blood is expelled, increasing the stroke volume and cardiac output.

At rest the ventricles do not completely empty with each beat; approximately 40% of the blood volume remains in the ventricles after contraction. During exercise stronger contractions occur, ejecting more of the residual blood from the heart. In Figure 7.33 you can see the effect of exercise on stroke volume.

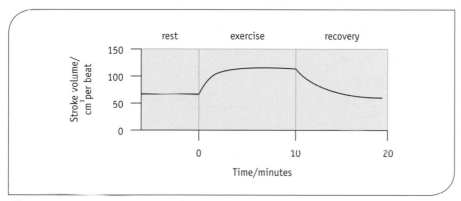

Figure 7.33 The effect of exercise on stroke volume.

Heart rate

Each of us has a slightly different resting heart rate. Measure your heart rate by counting your pulse while sitting at your desk. With each beat of your heart, a pulse of blood is ejected. This can be felt passing along the arteries; you can feel it fairly easily at the wrist (radial artery) or neck (carotid artery). The average heart rate for males is 70 beats per minute (bpm), while for females it is 72 bpm. The average fit person's heart rate is around 65 bpm.

Differences in resting heart rate are caused by many factors. For example, we differ in size of heart, owing to differences in body size and genetic factors. A larger heart will usually have a lower resting heart rate. It will expel more blood with each beat and so, other things being equal, does not have to beat as frequently to circulate the same volume of blood round the body. Endurance training produces a lower resting heart rate, largely due to an increase in the size of the heart, resulting from thickening of the heart muscle walls. The cyclist Miguel Indurain, five times winner of the Tour de France, had a resting pulse rate of 28 beats per minute.

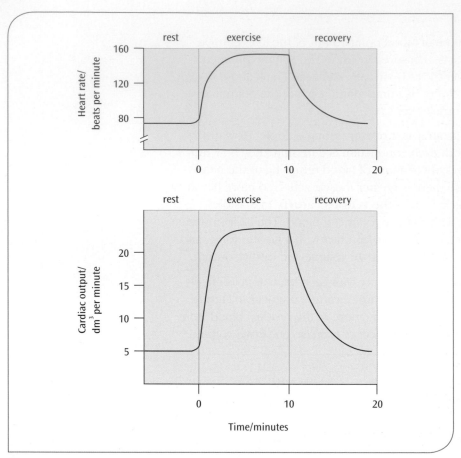

Figure 7.34 The effect of exercise on heart rate and cardiac output.

Look at the graphs in Figure 7.34 to see the effect of exercise on heart rate and cardiac output.

Q7.17 **a** A person has a resting stroke volume of 75 cm³. They take their pulse rate and find that it is 70 beats per minute. What is their cardiac output?
b An endurance athlete has the same cardiac output at rest, but has a resting heart rate of 50 bpm. What is their stroke volume?
c The cyclist Lance Armstrong, seven times winner of the Tour de France, has a resting heart rate of 33 bpm. What will his resting stroke volume be assuming his resting cardiac output is much the same as everyone else's?

Remind yourself about the cardiac cycle using Activity 1.6 in Topic 1.

The heart can beat even when it is removed from the body and placed in glucose and salt solution. This shows that the heart muscle is **myogenic**; it can contract without external nervous stimulation. The sinoatrial node in the wall of the right atrium produces stimulations which initiate the heartbeat.

However, the heart rate is under the control of the nervous system and is also affected by hormonal action.

Nervous control of heart rate

Heart rate is under the control of the **cardiovascular control centre** located in the medulla of the brain.

Nerves that form part of the **autonomic nervous system** (the nervous system over which you have no control) lead from the cardiovascular control centre to the heart. There are two such nerves going from the cardiovascular control centre to the heart – a **sympathetic nerve** (accelerator) and the **vagus nerve** which is a **parasympathetic nerve** and acts as a decelerator. See Table 7.2 for a comparison of the functions of these different nerve types. Stimulation of the sinoatrial node (SAN) by the sympathetic nerve causes an increase in the heart rate, whereas impulses from the vagus nerve slow down the rate (Figure 7.35).

The cardiovascular control centre detects accumulation of carbon dioxide and lactate in the blood, reduction of oxygen, and increased temperature. Mechanical activity in muscles and joints is detected by sensory receptors in muscles, and impulses are sent to the cardiovascular control centre. These changes result in higher heart rate.

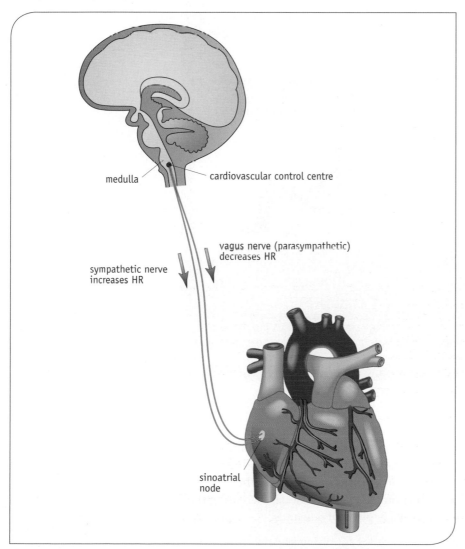

▲ **Figure 7.35** Control of the heart rate by the cardiovascular control centre.

At the sound of the starting pistol, or the sight of prey (an animal's next meal), skeletal muscles contract, and stretch receptors in the muscles and tendons are stimulated and send impulses to the cardiovascular control centre. This in turn raises the heart rate via the sympathetic (accelerator) nerve. There is an increase in venous return which leads to a rise in the stroke volume. Together, the elevated heart rate and stoke volume result in higher cardiac output, thus transporting oxygen and fuel to muscles more quickly.

Blood pressure rises with higher cardiac output. To prevent it rising too far, pressure receptors in the aorta and in the carotid artery send nerve impulses back to the cardiovascular control centre. Inhibitory nerve impulses are then sent from here to the sinoatrial node, so an excessive rise in blood pressure is prevented by negative feedback, preventing further rise in heart rate.

 Table 7.2 Sympathetic and parasympathetic nerves.

The autonomic (unconscious) part of your nervous system is made up of two sets of nerves:

- Sympathetic – stimulation of the sympathetic nerves prepares the body's systems for action (for the fight or flight response).

- Parasympathetic – stimulation of the parasympathetic nerves controls the body's systems when resting and digesting.

Examples of the opposing effects of the two types of nerve are shown below:

Organ or tissue	Effect of sympathetic stimulation	Effect of parasympathetic stimulation
intercostal muscles	increases breathing rate	decreases breathing rate
heart	increases heart rate and stroke volume	decreases heart rate and stroke volume
gut	inhibits peristalsis*	stimulates peristalsis

*Smooth muscle contractions in the gut wall that move food through the gut.

Hormonal effects on heart rate

Fear, excitement and shock cause the release of the hormone **adrenaline** into the bloodstream from the adrenal glands located above the kidneys. Adrenaline has a similar effect on the heart rate as stimulation by the sympathetic nerve. It has a direct effect on the sinoatrial node, increasing the heart rate to prepare the body for any likely physical demands.

Adrenaline also causes dilation of the arterioles supplying skeletal muscles, and constriction of arterioles going to the digestive system and other non-essential organs; this maximises blood flow to the active muscles. Before the start of a race, adrenaline causes an anticipatory increase in heart rate.

Q7.18   The carotid artery at the side of the neck is sometimes used to measure heart rate (pulse rate).
a Suggest why pressing on the carotid artery might reduce the pulse rate, thereby giving a false reading.
b Where could you take a pulse more reliably?

Q7.19   Look at Figure 7.34. Explain why the heart rate starts to rise before the start of exercise and suggest why this may be an advantage for the animal.

Breathing
Lung volumes

Figure 7.36 A spirometer can be used to measure lung volumes; changes in volume can be recorded as a trace.

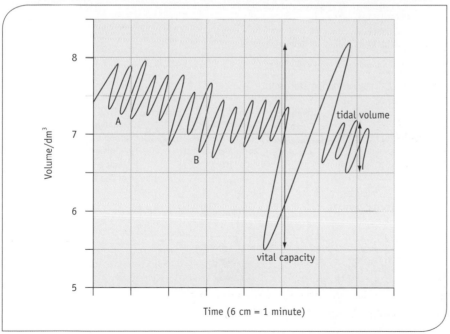

Figure 7.37 A spirometer trace showing quiet breathing with one maximum breath in and out. The trace can be used to measure depth and frequency of breathing. The fall in the trace is due to consumption of oxygen by the subject and the rate of oxygen consumption can be calculated by dividing the decrease in volume by time for the fall. Try analysing this trace in Question 7.21.

The volume of air we breathe in and out at each breath is our **tidal volume**. At rest this is usually about $0.5\,dm^3$. When exercise begins we increase our breathing rate and depth of breathing. The maximum volume of air we can inhale and exhale is our **vital capacity**. In most people this is 3–$4\,dm^3$, but in large or very fit people it can be $5\,dm^3$ or more. Singers and those playing wind instruments may also have a large vital capacity. Lung volumes including tidal volume and vital capacity can be measured using a spirometer (see Figure 7.36 and Activity 7.10).

The volume of air taken into the lungs in one minute, the **minute ventilation**, is calculated by multiplying the tidal volume (the average volume of one breath, in dm^3) by the breathing rate (number of breaths per minute).

Activity

In **Activity 7.10** you can measure lung volumes and breathing rate.
A7.10S

Q7.20 At rest, the average person takes 12 breaths per minute. Assuming a tidal volume of $0.5\,dm^3$, calculate the volume of air breathed in each minute.

Q7.21 From the spirometer trace in Figure 7.37:
a Read the tidal volume.
b Read the vital capacity.
c Work out the minute ventilation.
d Work out the volume of oxygen consumed between points A and B on the trace and calculate the rate of oxygen consumption.

The control of breathing

The **ventilation centre**, in the medulla oblongata of the brain, controls breathing; this is summarised in Figure 7.38.

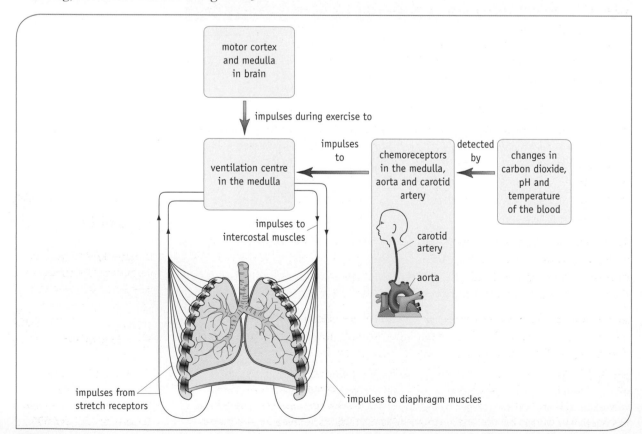

▲ **Figure 7.38** Control of breathing.

Inhalation

The ventilation centre sends nerve impulses every 2–3 seconds to the external intercostal muscles and diaphragm muscles. Both these sets of muscles contract, causing inhalation.

During deep inhalation, not only the external intercostals and diaphragm muscles but also the neck and upper chest muscles are brought into play.

Exhalation

As the lungs inflate, stretch receptors in the bronchioles are stimulated. The stretch receptors send inhibitory impulses back to the ventilation centre. As a consequence, impulses to the muscles stop and the muscles relax, stopping inhalation and allowing exhalation.

Exhalation occurs by the elastic recoil of the lungs (like a deflating balloon), and by gravity helping to lower the ribs. Not all of the air in the lungs is exhaled with each breath. The air remaining in the lungs, residual air, mixes with the air inhaled with each breath.

The internal intercostal muscles only contract during deep exhalation; for example during vigorous exercise a larger volume of air is exhaled, leaving less residual air in the lungs.

Controlling breathing rate and depth

At rest, the most important stimulus controlling the breathing rate and depth of breathing is the concentration of dissolved carbon dioxide in arterial blood via its effect on pH. A small increase in blood carbon dioxide concentration causes a large increase in ventilation. This is achieved in this way:

- Carbon dioxide dissolves in the blood plasma, making carbonic acid.
- Carbonic acid dissociates into hydrogen ions and hydrogencarbonate ions, thereby lowering the pH of the blood.

$$CO_2 + H_2O \rightleftharpoons H_2CO_3 \rightleftharpoons H^+ + HCO_3^-$$

- Chemoreceptors sensitive to hydrogen ions are located in the ventilation centre of the medulla oblongata. They detect the rise in hydrogen ion concentration.
- Impulses are sent to other parts of the ventilation centre.
- Impulses are sent from the ventilation centre to stimulate the muscles involved in breathing.

There are also chemoreceptors in the walls of the carotid artery and aorta which are stimulated by changes in pH resulting from changes in carbon dioxide concentration. These chemoreceptors monitor the blood before it reaches the brain and send impulses to the ventilation centre.

Increasing carbon dioxide concentration and the associated fall in pH leads to an increase in rate and depth of breathing, through more frequent and stronger contractions of the appropriate muscles. These more frequent and deeper breaths maintain a steep concentration gradient of carbon dioxide between the alveolar air and the blood (Figure 7.39). This in turn ensures efficient removal of carbon dioxide and uptake of oxygen. The opposite response occurs with a decrease in carbon dioxide. The control of carbon dioxide levels in the blood is an example of homeostasis operating via negative feedback.

Activity

In **Activity 7.11** you can investigate the control of ventilation rate. **A7.11S**

Extension

To find out how the blood carries oxygen and carbon dioxide read **Extension 7.3**. **X7.03S**

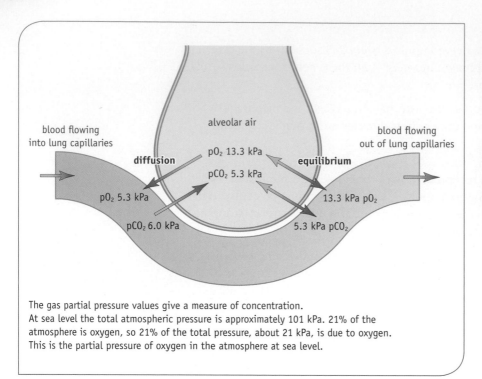

The gas partial pressure values give a measure of concentration.
At sea level the total atmospheric pressure is approximately 101 kPa. 21% of the atmosphere is oxygen, so 21% of the total pressure, about 21 kPa, is due to oxygen. This is the partial pressure of oxygen in the atmosphere at sea level.

🔺 **Figure 7.39** The exchange of carbon dioxide and oxygen between lung capillaries and alveoli.

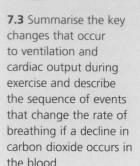

Controlling breathing during exercise

Immediately exercise begins, the impulses from the motor cortex of the brain, the region that controls movement, have a direct effect on the ventilation centre, increasing ventilation sharply. Ventilation is also increased in response to impulses reaching the medulla oblongata from stretch receptors in tendons and muscles involved in movement. The various chemoreceptors sensitive to carbon dioxide levels and to changes in blood temperature increase the depth and rate of breathing via the ventilation centre. The control mechanisms are summarised in Figure 7.38. There are receptors sensitive to changing oxygen concentrations in the blood; however they are rarely stimulated under normal circumstances.

Q7.22 During vigorous exercise, the concentration of oxygen in the lungs is higher than when at rest. Suggest the reasons for, and the advantage of, this elevated oxygen level.

Did you know? Altitude sickness

The control of breathing can go wrong if we climb too quickly to high altitude. As oxygen becomes scarce (at about 4000 m), the blood oxygen concentration can fall to a very low level. This triggers the medulla oblongata to make us breathe deeply and rapidly, especially if we are doing a hard climb. The hard panting flushes out too much carbon dioxide, resulting in a rise of blood pH. The ventilation centre may respond by stopping breathing altogether for a few seconds, so we alternate between heavy panting and not breathing at all. If this happens it is essential to descend quickly to a lower altitude.

Q7.23 Suggest why it is beneficial that stimulation of stretch receptors in the muscles increases ventilation.

Q7.24 When a person breathes air containing 80% oxygen, the minute ventilation is reduced by 20%. Explain how this occurs.

All muscle fibres are not the same

Aerobic capacity is dependent not only on uptake and transport of oxygen to the muscles, but also on efficiency of use once it reaches the muscle fibres. Although the muscles of all mammals are almost identical in their macroscopic and microscopic structure, it is possible to identify two distinct types of fibre.

If you eat chicken (which has the same types of muscle fibres as humans and other mammals) you may have noticed that some parts of the flesh are darker and others lighter. The breast meat, where the flight muscles are, is pale (the 'white meat') whereas the leg muscles are darker. This difference is caused by the fact that the two regions contain two different types of muscle fibre, which reflects the different functions of the muscles.

Slow-twitch fibres

Chickens spend most of their time on the ground, standing still or walking. The darker muscle in the legs and body is made up of fibres called **slow-twitch fibres**. These are specialised for slower, sustained contraction, and can cope with long periods of exercise. To do this they need to carry out a large amount of aerobic respiration.

Key features of slow-twitch muscle fibres

Slow-twitch muscle fibres have many mitochondria and high concentrations of respiratory enzymes to carry out aerobic reactions. They also contain large amounts of the dark red pigment **myoglobin**, which gives them their distinctive colour. Myoglobin is a protein similar to haemoglobin (the oxygen-carrying pigment found in red blood cells). It has a high affinity for oxygen and releases it only when the concentration of oxygen in the cell falls very low; it therefore acts as an oxygen store within muscle cells. Slow-twitch fibres are associated with numerous capillaries to ensure a good oxygen supply.

Fast-twitch fibres

In moments of panic or to reach higher perches chickens can perform near vertical take-offs, though they cannot fly for long. The paler flight muscle is largely made up of a different type of muscle fibre called **fast-twitch fibres**. These fibres are specialised to produce rapid, intense contractions. The ATP used in these contractions is produced almost entirely from anaerobic glycolysis.

Key features of fast-twitch muscle fibres

Fast-twitch fibres have few mitochondria. (Remember that glycolysis does not occur in mitochondria.) They also have very little myoglobin, so have few reserves of oxygen and few associated capillaries. With their reliance on anaerobic respiration there is a rapid build-up of lactate, so fast-twitch muscle fibres fatigue easily.

With aerobic training fast-twitch fibres can take on some of the characteristics of slow-twitch fibres, for example increased numbers of mitochondria, allowing them to use aerobic respiration reactions when contracting.

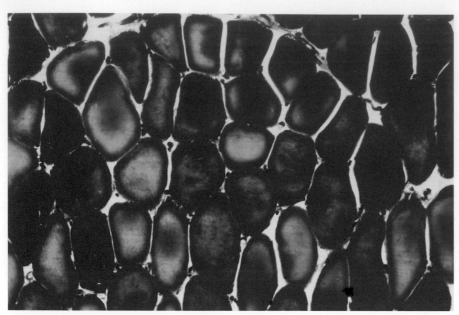

🔺 **Figure 7.40** Notice how some of the fibres are very dark. These contain more myoglobin and are slow-twitch fibres.

Q7.25 Study Table 7.3 and then answer the following questions.

a What is the significance of the large number of mitochondria in slow-twitch fibres?

b Why do fast-twitch fibres need more sarcoplasmic reticulum?

c Why will a large amount of myoglobin be advantageous to slow-twitch fibres?

d How will each type of fibre regenerate ATP?

e Which type of fibre will build up an oxygen debt more quickly? Give a reason for your answer.

🔻 **Table 7.3** Characteristics of the two muscle fibre types.

Slow-twitch fibres	Fast-twitch fibres
red (a lot of myoglobin)	white (little myoglobin)
many mitochondria	few mitochondria
little sarcoplasmic reticulum	extensive sarcoplasmic reticulum
low glycogen content	high glycogen content
numerous capillaries	few capillaries
fatigue resistant	fatigue quickly

What makes a sprinter?

In mammals, these two types of muscle fibre are not separated but found together in all skeletal muscles. The proportion of each type appears to be genetically determined, and varies between different people. Research has shown that successful endurance athletes such as marathon runners, rowers and cross-country skiers, with high aerobic capacity, have a higher proportion of slow-twitch muscle fibres in their skeletal muscles, up to 80%. In contrast, sprinters may have as few as 35% slow-twitch fibres. Throwers and jumpers have a more or less equal proportion of the two types.

Individuals may be better suited to a particular type of sport if they naturally have a higher proportion of fibres used in that activity. This, of course, will not be the only factor that contributes to an individual's success in a sport. For example, an individual with a highly efficient cardiovascular system will be well suited to aerobic exercise.

Q7.26 Which type of fibre might dominate the leg muscles of:
a the cheetah **b** the wildebeest?

7.4 Breaking out in a sweat

The maximum distance a cheetah can sprint is approximately 500 m. After this point, not only will the fast-twitch muscle fibres have fatigued due to the build-up of lactate, but the body temperature will have risen. Cheetahs must stop to recover.

The demands of physical activity can increase metabolic activity by up to 10 times, releasing energy which could potentially increase core temperature by around 1 °C every 5 to 10 minutes. This energy must be dispersed to maintain thermal balance.

The marathon, more than any other sport, has a history of heat-related deaths. In 490 BC Pheidippides, an intercity messenger of Ancient Greece, ran 26 miles from Marathon to Athens with instructions that the Athenians should not surrender to the Persian fleet. Legend has it that at the end of his journey Pheidippides dropped dead from exhaustion. In the 1912 Olympic Games in Stockholm, the Portuguese runner Lazaro collapsed from heat stroke after running 19 miles. He died the following day. Nowadays marathons tend to be scheduled in the early morning to reduce the chance of heat stroke. However, this is unpopular with TV schedulers and runners are sometimes required to race in more dangerous conditions to maximise advertising revenues. At the Athens Olympics in 2004 the marathons were run at peak viewing time in temperatures of 33 °C (men) and 35 °C (women) with humidity in excess of 40%.

▼ **Figure 7.41** During strenuous exercise enough heat is produced to raise our body temperature by 1 °C every 5 to 10 minutes were it not dissipated.

Q7.27 **a** What is the optimal temperature range for human cells?
b What will happen to metabolic reactions if body temperature falls below or rises above the normal range?

As we saw in Topic 6, core temperature is normally maintained within the narrow range of around 36–8 °C in humans using negative feedback. A rise of only 5 °C in core body temperature can be fatal.

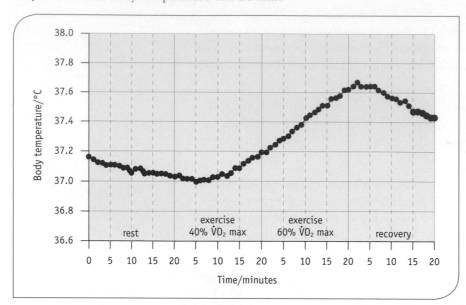

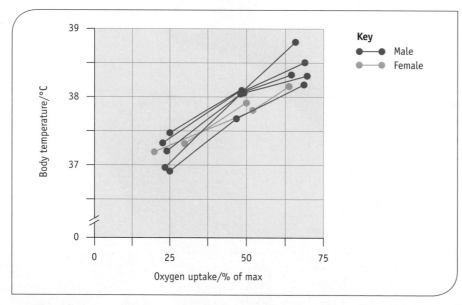

Figure 7.42 The effect of a 40-minute period of exercise on temperature (rectal measurements). *Source: NASA.*

Figure 7.43 Changes in core body temperature with increasing exercise intensity expressed as oxygen uptake.

During exercise, core body temperature rises (Figure 7.42) and this rise is related to intensity of exercise (Figure 7.43). The comparatively slight rise in temperature does not indicate a failure of the body to regulate the temperature; without mechanisms to redistribute energy though, the increase would quickly reach dangerous levels.

Temperature regulation

Thermoregulation during exercise is a good example of a homeostatic process; the body maintains a near-constant optimal internal temperature using negative feedback. The body must:

• detect changes in internal conditions using receptors

• coordinate the action of effectors to oppose the change and restore normal conditions.

Once the hypothalamus detects a deviation from the norm in core temperature, it starts a chain of actions which will counteract the deviation and bring body temperature back to the norm.

Use Figure 7.44 to remind yourself of the ways you can raise and lower body temperature.

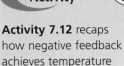

Activity

Activity 7.12 recaps how negative feedback achieves temperature regulation. **A7.12S**

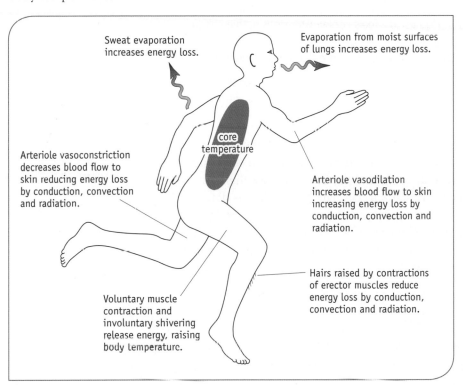

Sweat evaporation increases energy loss.

Evaporation from moist surfaces of lungs increases energy loss.

core temperature

Arteriole vasoconstriction decreases blood flow to skin reducing energy loss by conduction, convection and radiation.

Arteriole vasodilation increases blood flow to skin increasing energy loss by conduction, convection and radiation.

Hairs raised by contractions of erector muscles reduce energy loss by conduction, convection and radiation.

Voluntary muscle contraction and involuntary shivering release energy, raising body temperature.

▲ **Figure 7.44** How energy is transferred to and from the body.

Methods of energy transfer

• Radiation – energy can be radiated from one object to another through air, or even a vacuum, as electromagnetic radiation. Our bodies are usually warmer than the surrounding environment, so we radiate energy. Of course, this can operate in reverse – even in sub-freezing conditions such as skiing in the mountains, a person can remain warm due to the heat energy radiated directly from the Sun, or reflected back from the snow.

• Conduction – energy loss by conduction involves direct contact between objects and energy transfer from one to another. Sitting on a cold rock will cause body cooling, as energy is transferred to the rock by conduction.

• Convection – air lying next to the skin will be warmed by the body (unless the air temperature exceeds body temperature). If the air expands and rises, or is moved away by air currents, it will be replaced by cooler air which can then also be warmed by the body. This is energy loss by convection. Trapping a layer of still air next to the skin using fur or thermal underwear is an effective method of thermal insulation.

- Evaporation – energy is needed (latent heat of evaporation) to convert water from liquid to vapour. It requires 2400 kJ of energy to evaporate 1 dm^3 of water. The energy required to evaporate sweat is drawn from the body, cooling it. However, sweating is only effective if the water actually evaporates from the surface of the skin. In conditions of high humidity this becomes virtually impossible, and controlling body temperature becomes much more difficult. Mammals, birds and reptiles may also pant to keep cool, by evaporation of water from the gas exchange surfaces.

Q7.28 **a** The marathon runner in Figure 7.41 feels cooler after dousing himself in water. How does this help to cool him?
b Why does high humidity conditions make marathon running more dangerous?

Q7.29 Cats (including the big cats) can only sweat from the skin surface of their paws and nose. What other method might a cheetah rely on for transfer of energy to the environment?

At 37°C, human core body temperature is normally higher than the surroundings, so energy will be transferred to the environment. In very cold environments, excessive cooling may occur and the body can lose thermal balance – the core body temperature starts to fall. The hypothalamus detects this and immediately does its best to regulate the internal temperature by increasing metabolic rate and slowing energy loss, as outlined in Topic 6 and Figure 7.44. Although less common during exercise, there are occasions when the body faces this challenge (see Figure 7.45).

Q7.30 **a** What will be the major method of energy loss for the cross-Channel swimmer in Figure 7.45?
b Swimming the English Channel in summer means spending between 10 and 20 hours in water that is usually between 13 and 16°C. What mechanisms are used to maintain body temperature and enable survival without wearing a wetsuit?

▲ **Figure 7.45** Channel swimming exposes the body to cold stress.

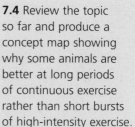

Checkpoint

7.4 Review the topic so far and produce a concept map showing why some animals are better at long periods of continuous exercise rather than short bursts of high-intensity exercise.

7.5 Overdoing it

Although we will never run as fast as a sprinting cheetah, over the years top athletes have been getting faster. At the end of the 1920s the 100 m world records stood at 10.4 and 12.0 seconds for men and women respectively. Since then, the times have steadily fallen, and at the start of the twenty-first century the records stood at 9.79 and 10.49 seconds. This improvement in performance has been achieved through more frequent and targeted training, improved nutrition and advances in the design and materials used in athletes' clothing, footwear and tracks.

But some athletes attempt to do more than their body can physically tolerate and reach the point where they have inadequate rest to allow for recovery (Figure 7.46). This is known as overtraining. 'Burnout' symptoms due to overtraining can persist for weeks or months. The symptoms are varied and, in addition to poor athletic performance and chronic fatigue, can include immune suppression leading to more frequent infections and increased wear and tear on joints, which may require surgical repair.

Excessive exercise and immune suppression

Athletes engaged in heavy training programmes seem more prone to infection than normal. Sore throats and flu-like symptoms (upper respiratory tract infections, URTIs) are more common. Some scientists have suggested there is a U-shaped relationship between risk of infection and amount of exercise (Figure 7.47).

Figure 7.46 Top sportspeople can suffer from overtraining. At the end of the season some professional footballers have been found to have very low white blood cell counts.

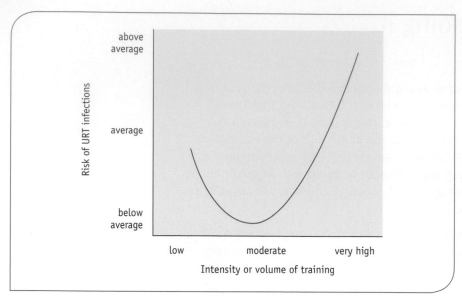

Figure 7.47 Moderate exercise seems to offer some protection against upper respiratory tract infections.

Q7.31 What do these data suggest about the risk of upper respiratory tract infections related to exercise?

In a study of participants in the Los Angeles marathon, it was found that 13% of the runners reported upper respiratory tract infections in the week after the race. A control group of runners of comparable fitness who had not taken part had an infection rate in the week after the race of only 2%. There is much discussion about whether this is a true cause and effect relationship.

Two main factors have been suggested as contributing to higher infection rates: increased exposure to pathogens, and suppressed immunity with hard exercise.

The location of the competition and any necessary travel may expose the athlete to a greater range of infected people and unfamiliar microorganisms. This alone could increase the occurrence of infection, even if overtraining did not suppress their immunity. Participation in team sports will also bring players into close contact with others and increase the chances of transmission of infection.

Effects of exercise on immunity

A number of research studies have shown that different components of both the non-specific and specific immune systems are affected to various degrees by both moderate and excessive exercise.

Moderate exercise

Moderate exercise increases the number and activity of a type of lymphocyte called **natural killer cells**. Natural killer cells are found in the blood and lymph; unlike the B and T cells, they do not use specific antigen recognition. They provide non-specific immunity against viruses and other intracellular microbes or cancerous cells. They recognise glycoproteins on the surface of the target cell and secrete apoptosis-inducing molecules, causing lysis of the cells. They thus offer non-specific protection against upper respiratory tract infections and other infections.

Vigorous exercise

Research shows that during recovery after vigorous exercise, the number and activity of some cells in the immune system fall. These include:

- natural killer cells
- phagocytes
- B cells
- T helper cells.

The specific immune response is depressed as a consequence of these changes.

The decrease in T helper cells reduces the amount of cytokines available to activate T and B cells. This in turn reduces the quantity of antibody produced. It has also been suggested that an inflammatory response occurs in muscles due to damage to muscle fibres caused by heavy exercise, and this may reduce the available non-specific immune response against upper respiratory tract infections.

Q7.32 **a** Which of the white blood cells mentioned above secretes antibodies?
b What are the main features of the non-specific immune response to infection?
c How will the action of T killer cells be affected by the decrease in T helper cell numbers?

There is also much debate as to whether the effects of intense exercise are caused by the activity itself or by related psychological stress due to heavy training schedules and competition. Both physical exercise and psychological stress cause secretion of hormones such as adrenaline and cortisol (a hormone also secreted by the adrenal glands), both of which are known to suppress the immune system.

Q7.33 How does the above evidence support the idea that moderate exercise enhances immunity while excessive exercise suppresses it?

> **Activity**
>
> In **Activity 7.13** you can summarise your knowledge of the immune system and immune suppression. **A7.13S**

How are joints damaged by exercise?

Professional athletes, such as football, hockey and rugby players, risk developing joint injuries due to the high forces the sport generates on their joints. Repeated forces on joints such as the knee can lead to wear and tear of one or more parts of the joint. Many joint disorders are associated with such overuse, some of which can also result from ageing. These disorders are typically associated with pain, inflammation and restricted movement of the joint. Treatment usually involves rest, ice, compression and elevation (RICE), anti-inflammatory painkillers, and, if necessary, surgical repair.

Knees are particularly susceptible to wear and tear injuries. The problems include:

- The articular cartilage covering the surfaces of the bones wears away so that the bones may actually grind on each other, causing damage which can lead to inflammation and a form of arthritis.
- Patellar tendinitis (jumper's knee) occurs when the kneecap (patella) does not glide smoothly across the femur due to damage of the articular cartilage on the femur.

- The bursae (fluid sacs) which cushion the points of contact between bones, tendons and ligaments can swell up with extra fluid. As a result, they may push against other tissues in the joint, causing inflammation and tenderness. Bursitis of the knee is also known as 'housemaid's knee' because it was common in housemaids due to the repetitive kneeling associated with their work.
- Sudden twisting or abrupt movements of the knee joint often result in damage to the ligaments.

How can medical technology help?

Improvements in medical technology over recent years, including the development of prosthetic limbs and keyhole surgery procedures, have enabled the disabled and those with injuries to participate in sport.

Keyhole surgery

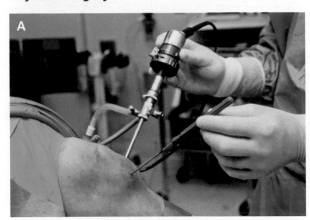
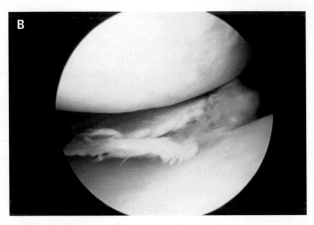

🔺 **Figure 7.48** Arthroscopy (**A**) allows surgeons to see within joints and repair damage, such as the degenerative tear in the pad of cartilage (meniscus) of this knee joint (**B**). There are two menisci in each knee. These crescent-shaped pieces of cartilage reduce friction in the moving joint and increase its stability.

Injuries to joints can limit the amount of exercise a person can take and have often shortened the careers of professional athletes. Surgical operations to repair damage used to be painful and recovery took a long time. The main reason for this was the large incision needed to remove or repair even very small structures. A large hole had to be made to give access with space for the surgeon's hands and instruments, and also let in enough light to allow the surgeon to see what they were doing. These large incisions caused a good deal of bleeding, a lot of pain, increased risk of infection, and prolonged recovery after the operation.

With the advent of **keyhole surgery**, using fibre optics or minute video cameras, all this has changed. It is now possible to repair damaged joints or remove diseased organs through small holes. Keyhole surgery on joints is known as **arthroscopy** (Figure 7.48A). This literally means 'to look within the joint'. To carry out an arthroscopic examination, the surgeon makes only a very small incision into the skin of the patient. Then small surgical instruments about the size of a pencil, along with a fibre-optic tube to view the inside of the joint, are placed inside the joint. A TV camera and light source are attached to the fibre-optic tube, to allow the surgeon to see inside the joint (Figure 7.48B). As a result of making a very small incision, recovery is rapid, and only a short stay in hospital is needed.

Although the inside of most joints can be viewed with an arthroscope, the six joints most frequently examined and treated are the knee, shoulder, elbow, ankle, hip and wrist. As advances are made in medical technology and new techniques are developed by surgeons, other joints may be treated more frequently in the future.

Prostheses

A **prosthesis** (plural prostheses) is an external artificial body part used by someone with a disability to enable him or her to regain some degree of normal function or appearance. By using specialised prostheses, disabled athletes can be more physically active and perform at higher levels (Figure 7.49).

Figure 7.49 Two different types of leg prosthesis in use by athletes. The female athlete is wearing two prostheses that are specially designed for sprinting and used only for this purpose. The male athlete is wearing his normal prosthesis. Can you tell which leg it is?

There have been significant developments in prosthetic limbs over recent years, with the introduction of variations in design for different activities. For example, athletes with prosthetic legs may use a dynamic response prosthetic foot. Such a foot changes its shape under body weight but returns to the original shape on lifting off the ground. This puts spring into the step and provides sure footing.

Prosthetic feet may also be articulated (have joints) or not. Articulated feet are better on uneven surfaces, and so useful in sports such as golf. In some sports, a foot may not be required at all; a flipper could be used for swimming, or a pedal-binding for cycling. High-friction surfaces can be added to provide better grip to prostheses used for rock climbing.

Activity

Check out the video on bone damage and repair in **Activity 7.14**. **A7.14S**

Doing too little exercise

Although overuse can result in damage to joints and bones there are, of course, advantages to doing physical activity. These include:

- lowered blood pressure due to increased arterial vasodilation, which reduces the risk of coronary heart disease and stroke
- increased level of blood HDLs which transport cholesterol to the liver where it is broken down, and reduced levels of LDLs which are associated with development of atherosclerosis in coronary heart disease and stroke (see Topic 1, lipoproteins)
- maintaining a healthy weight by helping to achieve a balance between energy input and output
- increased sensitivity of muscle cells to insulin which improves blood glucose regulation and reduces the likelihood of developing type II diabetes
- increasing bone density and slowing its loss during old age, which delays and reduces the progress of the bone-wasting disease osteoporosis
- reduced risks of some cancers
- improved mental well-being.

The Department of Health's advice on physical activity recommends that adults take 30 minutes of at least moderate activity on at least five days a week. Physical activity includes walking, cycling, gardening, and active hobbies or sports. Doing too little exercise means that you do not gain the health benefits described above and you are at a higher risk of coronary heart disease, stroke, cancer, obesity, diabetes and osteoporosis.

A sedentary lifestyle combined with overeating and drinking can lead to weight gain and potentially to obesity. A person is considered obese if their body mass index is over 30 (BMI = body mass in kg/(height in m)^2; see Topic 1). Obesity leads to high blood pressure and high blood LDL levels which increase the risk of coronary heart disease and stroke.

Obesity also increases the risk of developing type II diabetes (often known as non-insulin-dependent diabetes or late-onset diabetes). High blood glucose levels due to eating sugar-rich foods can reduce sensitivity of cells to insulin and result in type II diabetes. The body does not respond to any insulin produced so cannot control blood sugar levels. There is decreased absorption of glucose from the blood; cells break down fatty acids and proteins instead, which leads to weight loss.

Weblink

Find out more about osteoporosis from the National Osteoporosis Society website, and diabetes from the Diabetes UK website.

Checkpoint

7.5 Produce an annotated list of the disadvantages of exercising **a** too much and **b** too little.

7.6 Improving on nature

Performance-enhancing substances

The use of drugs to enhance performance in sport is known as doping. It is thought that this term originates from the South African word 'doop', which referred to an alcoholic stimulant drink used in certain tribal ceremonies.

Doping is not just a recent problem; throughout history, some athletes have sought a competitive edge by the use of chemicals. As long ago as the third century BC, certain Ancient Greeks were known to ingest hallucinogenic mushrooms in an attempt to improve athletic performance (Figure 7.50). Roman gladiators used stimulants in the Circus Maximus to overcome fatigue and injury, while other athletes experimented with caffeine, alcohol and opium. The important social status of sport, and the high economic value of victory, both then and now, has placed great pressure on athletes to be the best. This pressure has increased the abuse of performance-enhancing drugs.

A wide range of substances can be taken for their performance-enhancing effects. Three examples are **erythropoietin**, **testosterone** and **creatine**. Erythropoietin and testosterone are currently banned whereas creatine is not.

Figure 7.50 Ancient drug cheats? At the Athens Olympics of 2004, 24 doping violations were discovered, the highest number for any Olympic games.

Erythropoietin

Erythropoietin (EPO) is a peptide hormone produced naturally by the kidneys. It stimulates the formation of new red blood cells in bone marrow. EPO can be produced using DNA technology, and is used to treat anaemia. As it is a natural substance, it has been difficult to test to see whether raised EPO levels are natural or not.

Q7.34 Explain how taking EPO would increase the performance of an endurance athlete.

There are health risks associated with this substance. If EPO levels are too high, the body will produce too many red blood cells, which can increase the risk of thrombosis, possibly leading to heart attack and stroke. Injections of EPO have been implicated in the deaths of several athletes.

French scientists have recently developed a technique capable of distinguishing between synthetic and natural EPO.

Q7.35 What is meant by thrombosis?

Q7.36 Explain why EPO is not taken by sprint athletes.

Q7.37 Why would it be important to distinguish between natural and recombinant EPO (synthetic EPO)?

Did you know? Banned substances

◀ **Figure 7.51** Alain Baxter, the British skier, was stripped of his bronze medal at the 2002 Winter Olympics for testing positive for a banned substance after using a nasal inhaler.

WADA, the World Anti-Doping Agency, promotes and coordinates the international fight against doping in sport. WADA aims to reinforce the 'ethical principles for the practice of doping-free sport and to help protect the health of athletes'.

The Olympic Movement Anti-Doping Code prohibits various substances and practices. The following classes of substances are prohibited:

- A: stimulants (drugs that increase heart rate and alertness) such as amphetamines, cocaine and excessive amounts of pseudoephedrine (a drug commonly found in over-the-counter decongestants) and caffeine (found in coffee, tea and chocolate)
- B: narcotics (powerful painkillers causing drowsiness) such as diamorphine (heroin), methadone, morphine and pethidine

- C: anabolic agents, including anabolic steroids such as testosterone and nandrolone, and beta-2 agonists such as salbutamol, a bronchodilator used in asthma inhalers
- D: diuretics (drugs that promote formation of urine) such as frusemide
- E: peptide hormones, mimetics and analogues, including EPO (erythropoietin) and insulin.

Competition athletes with medical conditions requiring prescription of these drugs need to obtain permission to use them. It is the responsibility of the athlete to check whether a drug is banned or not, and to ensure that he or she does not inadvertently take the banned substance (Figure 7.51).

Testosterone

Testosterone is a steroid hormone (made from cholesterol) produced in the testes by males and in small amounts by the adrenal glands in both males and females. Testosterone is one of a group of male hormones known as androgens, from the Greek *andros* meaning male or man.

Key biological principle: Hormones

Hormones are chemical messengers released from endocrine glands directly into the blood. (Unlike exocrine glands, such as sweat glands and salivary glands, endocrine glands do not have ducts.) Hormones are carried around the body and enter cells or bind to complementary receptor molecules on the cell membranes of specific target cells (see Topic 3, signal proteins). Each hormone has a characteristic response, resulting from its effect on enzymes (Figure 7.52).

Peptide hormones are protein chains, varying from about 10 to 300 amino acids in length. Peptide hormones cannot easily pass through a cell membrane because they are charged molecules. The binding of a peptide hormone to a receptor in the cell membrane activates another molecule in the cell cytoplasm which brings about chemical changes in the cell. Peptide hormones include EPO, human growth hormone and insulin.

Steroid hormones are formed from lipids, and have complex ring structures. Testosterone is a steroid hormone. Steroid hormones pass through the cell membrane and bind to receptor molecules within the cell which, once activated, function as transcription factors, switching enzyme synthesis on or off.

Q7.38 Why can steroid hormones pass through the cell membrane?

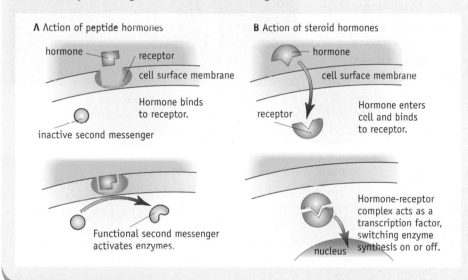

A Action of peptide hormones

hormone — receptor

cell surface membrane

Hormone binds to receptor.

inactive second messenger

Functional second messenger activates enzymes.

B Action of steroid hormones

hormone

cell surface membrane

receptor

Hormone enters cell and binds to receptor.

Hormone-receptor complex acts as a transcription factor, switching enzyme synthesis on or off.

nucleus

Figure 7.52 Hormones affect target cells by binding to receptors which control the production of enzymes.

Testosterone causes the development of the male sexual organs. During adolescence it is responsible for development of the male secondary sexual characteristics, for example the deepening of the voice, growth of facial and body hair, and skeletal and muscular changes. Character changes such as increased aggressiveness have been attributed to testosterone.

Testosterone binds to androgen receptors which are numerous on cells in target tissues. They modify gene expression to alter the development of the cell; for example they will increase anabolic reactions such as protein synthesis in muscle cells, increasing the size and strength of the muscle.

Athletes and bodybuilders (Figure 7.53) may use injections of testosterone to increase muscle development, but this is not very effective as testosterone is quickly broken down. To overcome this problem, synthetic **anabolic steroids** such as nandrolone have been manufactured by chemical modification of testosterone.

Medical experts see significant dangers in the use – and particularly the gross overuse – of anabolic steroids. For example, anabolic steroids can cause high blood pressure, liver damage, changes in the menstrual cycle in women, decreased sperm production and impotence in men, kidney failure and heart disease. They can increase aggression in both men and women. In women the androgenic (masculinising) side effects are not generally thought to be desirable.

Originally developed for the treatment of muscle-wasting diseases, anabolic steroids are also used in the treatment of osteoporosis. In the UK they are prescription-only drugs. They are classified as Class C drugs under the Misuse of Drugs Act, with the maximum penalty for the illegal *possession* of steroids currently (2005) standing at two years' imprisonment and/or a fine. *Supplying* a Class C drug, such as an anabolic steroid, can lead to heavier penalties, even if no money has changed hands. The International Olympic Committee has banned the use of anabolic steroids. The illegal use of steroids occurs not only in human sport but also in animal sports such as horse- and dog-racing.

⬤ **Figure 7.53** Anabolic steroids are used to increase muscle development. Heavy weight or resistance training is necessary for anabolic steroids to exert any beneficial effect on performance.

Anabolic steroids and their by-products can be detected relatively easily in urine samples by the technique of mass spectrometry. As these substances occur naturally it is difficult to set a level above which an athlete can confidently be said to be doping. Testosterone and a related compound, epitestosterone, are both found in urine. When an athlete takes an anabolic steroid, the ratio of testosterone to epitestosterone (the T/E ratio) increases. The International Olympic Committee states that an athlete with a T/E ratio above 6 is guilty of doping.

Creatine

Many athletes take dietary supplements containing an amino-acid derived compound known as creatine. Creatine is naturally found in meat and fish. Once ingested it is absorbed unchanged and carried in the blood to tissues such as skeletal muscle. Creatine is also synthesised in the body from the amino acids glycine and arginine. Creatine supplements have been reported to increase the amounts of creatine phosphate (CP) in muscles. The theoretical benefit of increased CP storage is an improvement in performance during repeated short-duration, high-intensity exercise. Research has shown improvements in activities such as sprinting, swimming and rowing. The use of creatine supplements combined with heavy weight training has been associated with increases in muscle mass and maximal strength, and a decrease in recovery time.

Creatine is considered as a nutritional supplement (Figure 7.54) and is therefore not on the list of prohibited substances. Some adverse effects of taking creatine supplements have been reported. These include diarrhoea, nausea, vomiting, high blood pressure, kidney damage and muscle cramps.

Activity

In **Activity 7.15** you can interpret data on the effects of testosterone.
A7.15S

Checkpoint

7.6 Outline the uses and misuses of the drugs creatine, testosterone and erythropoietin. Suggest ethical arguments for and against the use of drugs to improve sporting performance.

 Figure 7.54 Creatine supplements can be bought in high street stores.

Should performance-enhancing substance use be banned?

The pressure to succeed in competitive sport is ever increasing, not only due to the desire to be the best but also for the financial rewards and greater media interest. The desire to win combined with pressure and expectations from coaches, sponsors and the general public can be such that some athletes are prepared to take drugs that will enhance their performance, even if there are associated risks.

The International Olympic Committee (IOC) and other sporting bodies consider that the use of performance-enhancing substances is unhealthy and against the ethics of sport. The IOC started drug testing in 1968 after a Tour de France cyclist died from an amphetamine overdose; random testing began in 1989. The ban aims to protect the health of athletes and ensure that there is fair competition.

There are some people who consider that the use of substances is ethically acceptable, arguing that athletes have a right to decide whether they take the drug or not, deciding for themselves if the potential benefit is worth the risk to their health. Those who oppose this view may say that frequently the athletes do not make a properly informed decision, lacking information about the possible health consequences, and coming under pressure from others to take illegal drugs.

The idea that drug-free sport is fair is disputed by those who maintain that drug use is acceptable on the grounds that there is already inequality of competition due to the differences in time available for training and in resources. Some individuals may know that drug use is against the rules both of the governing bodies of sport and the idea of fair play but they are unwilling to be at a competitive disadvantage so choose not to adhere to the rules.

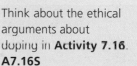 **Activity**

Think about the ethical arguments about doping in **Activity 7.16**. **A7.16S**

Activity

Use **Activity 7.17** to check your notes using the topic summary provided. **A7.17S**

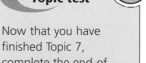 **Topic test**

Now that you have finished Topic 7, complete the end-of-topic test before starting Topic 8.

Topic ⑧ Grey matter

Why a topic called Grey matter?

The brain, with over 10^{11} neurones (nerve cells), is the most complicated organ in the body and makes even our most sophisticated computer seem simple. The brain influences our every sensation, emotion, thought, memory and action. At each moment of every day it is bombarded with sensory information from the world around us, and interprets this information to create a meaningful view of the world. Looking at the world is not merely inspecting a simple picture, like observing a slide show projected on a screen inside our heads. The information is processed to provide us with our experience of the world.

But sometimes things may not be as they seem, as the anthropologist Colin Turnbull found in the 1950s when he was working in what is now called the Democratic Republic of Congo. He and Kenge, one of the Bambuti people used to living in dense forest with only small clearings, went out onto the grassland plains of the former Congo. Looking across the wide open plain, Kenge turned to Turnbull and asked 'What insects are those?' (Figure 8.1). When Turnbull told Kenge that the insects were buffalo, he roared with laughter and told him not to tell such stupid lies.

🔺 **Figure 8.1** Kenge thought the buffalo in the distance were insects.

How does the nervous system function to let any of us look across a plain? Why did Kenge, an intelligent man, get the wrong impression? Was his visual development faulty or was he misinterpreting what he was viewing?

It is not just Kenge who is sometimes mistaken by what is viewed. Have a look at Figure 8.2 and decide which of the lines is longer. Now measure them to find out if you were correct. Why do many of us get this wrong when Zulu people are not fooled so easily?

All in the synapses

With upwards of 10^{14} interconnections between its neurones, the working of the brain is dependent on its synapses and their neurotransmitters. How do these function and how can they go wrong? Imbalances in naturally occurring brain chemicals and drugs that can cross the blood-brain barrier can affect synapses and have adverse consequences for health. How are synapses affected by conditions such as depression and Parkinson's disease and by the use of MDMA (ecstasy)?

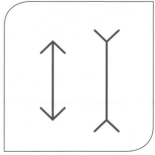

△ **Figure 8.2** Which line is longer?

Overview of the biological principles covered in this topic

In this topic you will revisit ideas about receptors and effectors introduced in Topics 6 and 7 by examining the pupil light reflex. Building on these ideas you will consider in more detail how stimuli are detected, as shown by receptor cells in the retina detecting light. You will also learn how the brain and eyes combine to enable visual perception. This leads to a discussion of the transmission of nerve impulses along axons and across synapses, and then to contrasting nervous and hormonal coordination.

You will look at how some diseases and drugs affect the brain to illustrate how chemicals can affect synaptic transmission. You will revisit genetic inheritance studied in Topics 2 and 5 and see how many mental disorders show polygenic inheritance. With reference to some of the DNA techniques covered earlier in the course you will discover how genetics can play a part in increasing our knowledge of brain structure and activity.

You will investigate the structures of the different regions of the brain and the evidence that links structure and function, including the use of imaging techniques you met in Topic 1. You will look at visual development and in particular the need for stimulation of synapses and the role of synapses in learning. Throughout this topic the contribution of nature and nurture to brain development is highlighted.

You will have the opportunity to discuss the ethical issues related to the use of animals in research.

8.1 The nervous system and nerve impulses

As Kenge and Colin Turnbull emerged from the forest, how did their eyes and brains work to let them look across the plain? Seeing is possible because the cells of the nervous system (Figure 8.3) are able to conduct **nerve impulses** and pass them to one another. In fact all our senses, emotions, memories and thoughts are dependent on nerve impulses.

The nervous system is highly organised (Figure 8.4), receiving, processing and sending out information, as we saw with temperature control in Topic 6 and control of heart rate in Topic 7.

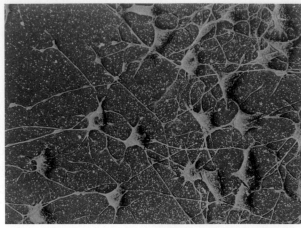

▲ **Figure 8.3** Nerve cells such as these form the basis of the nervous system.

What are nerve cells like?

It is important to distinguish between a **neurone**, which is a single cell, and a nerve. A **nerve** is a more complex structure containing a bundle of the axons of many neurones surrounded by a protective covering.

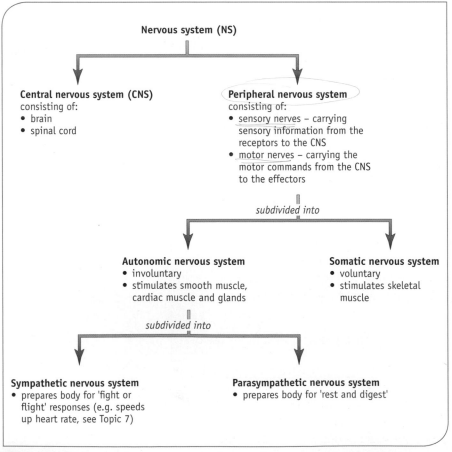

Nervous system (NS)

Central nervous system (CNS)
consisting of:
- brain
- spinal cord

Peripheral nervous system
consisting of:
- sensory nerves – carrying sensory information from the receptors to the CNS
- motor nerves – carrying the motor commands from the CNS to the effectors

subdivided into

Autonomic nervous system
- involuntary
- stimulates smooth muscle, cardiac muscle and glands

Somatic nervous system
- voluntary
- stimulates skeletal muscle

subdivided into

Sympathetic nervous system
- prepares body for 'fight or flight' responses (e.g. speeds up heart rate, see Topic 7)

Parasympathetic nervous system
- prepares body for 'rest and digest'

▲ **Figure 8.4** The organisation of the nervous system.

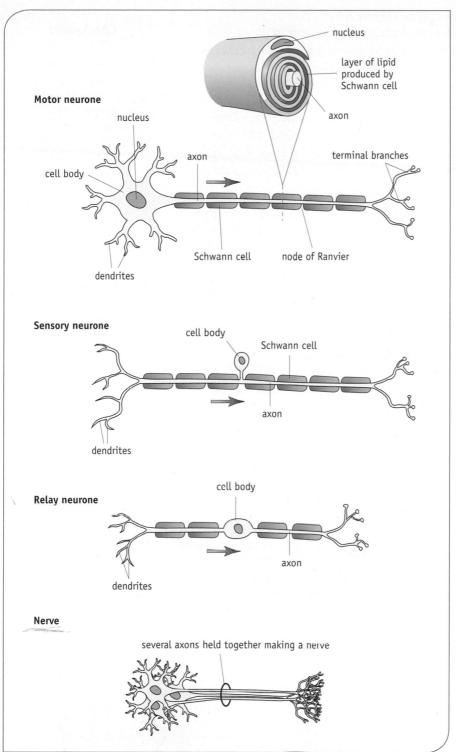

Figure 8.5 The structure of neurones.

Although there are different types of neurone, they all have the same basic characteristics. The **cell body** contains the nucleus and cell organelles within the cytoplasm. There are two types of thin extension from the cell body:

- very fine **dendrites** conduct impulses *towards* the cell body
- a single long process, the **axon**, transmits impulses *away from* the cell body.

There are three main types of neurone (Figure 8.5):

- **Motor neurones** – the cell body is always situated within the central nervous system (CNS) and the axon extends out, conducting impulses from the CNS to effectors (muscles or glands). The axons of some motor neurones can be extremely long, such as those that run the full length of the leg. Motor neurones are also known as effector neurones.
- **Sensory neurones** – these carry impulses from sensory cells to the CNS.
- **Relay neurones** – these are found mostly within the CNS. They can have a large number of connections with other nerve cells (Figure 8.3). Relay neurones are also known as connector neurones and as interneurones.

There is usually a fatty insulating layer called the **myelin sheath** around the axon. This is made up of **Schwann cells** wrapped around the axon (Figure 8.5). As we shall see later, the sheath affects how quickly nerve impulses pass along the axon (page 194). Not all animals have myelinated axons – they are not found in invertebrates, and some vertebrate axons are unmyelinated.

> **Checkpoint** ✓
>
> **8.1** Draw up a table comparing the structure and location of motor, relay and sensory neurones.

Reflex arcs

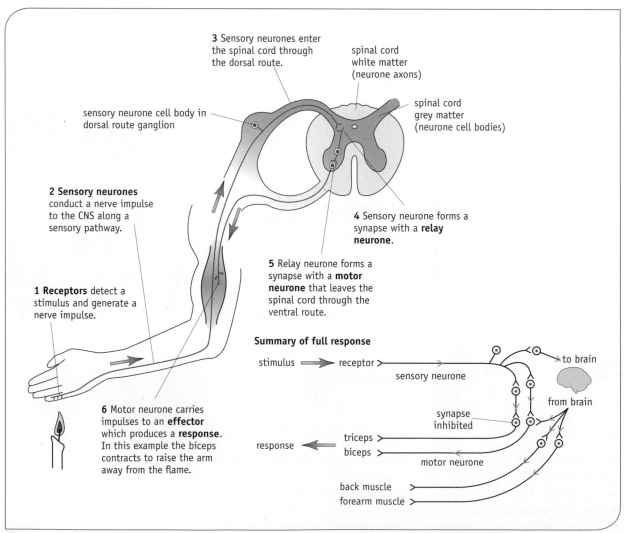

3 Sensory neurones enter the spinal cord through the dorsal route.

spinal cord white matter (neurone axons)

sensory neurone cell body in dorsal route ganglion

spinal cord grey matter (neurone cell bodies)

2 Sensory neurones conduct a nerve impulse to the CNS along a sensory pathway.

4 Sensory neurone forms a synapse with a **relay neurone**.

5 Relay neurone forms a synapse with a **motor neurone** that leaves the spinal cord through the ventral route.

1 Receptors detect a stimulus and generate a nerve impulse.

Summary of full response

stimulus ➡ receptor ⟩
sensory neurone
to brain
from brain

synapse inhibited

response ⬅ triceps ⟩
biceps ⟩
motor neurone
back muscle ⟩
forearm muscle ⟩

6 Motor neurone carries impulses to an **effector** which produces a **response**. In this example the biceps contracts to raise the arm away from the flame.

🔺 **Figure 8.6** A reflex arc allowing withdrawal of the hand.

Nerve impulses follow routes or pathways through the nervous system. Some pathways are relatively simple – the knee-jerk reflex involves just two neurones, a sensory neurone communicating with a motor neurone to connect receptor cells with effector cells. These simple nerve pathways are known as **reflex arcs** and are responsible for our **reflexes** – rapid, involuntary responses to stimuli. Figure 8.6 illustrates one reflex arc involving just three neurones.

However, most nerve pathways involve numerous neurones within the central nervous system. A sensory neurone connects to a range of neurones within the CNS and these pass impulses to the brain to produce a coordinated response. Figure 8.6 shows how even in reflex arcs there are additional connections within the CNS to ensure a coordinated response. Some synapses with motor neurones will be inhibited to ensure that the desired response occurs.

Q8.1 What is the advantage of reflex pathways?

Q8.2 Describe the nerve pathways involved if instead of a candle the hand was picking up a hot dinner plate, but the person did not want to drop the plate.

The pupil reflex

When Kenge and his companion emerged from the trees they moved from deep shade to bright sunlight. Immediately a reflex arc caused a change in the diameter of their pupils. If you cover your eyes for a few minutes, and then uncover your eyes while looking in a mirror, you can see that the size of your pupils decreases and the size of the irises increases (Figure 8.7).

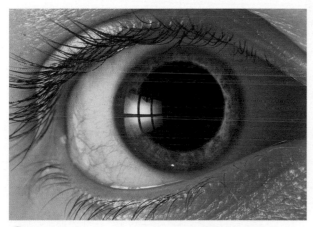 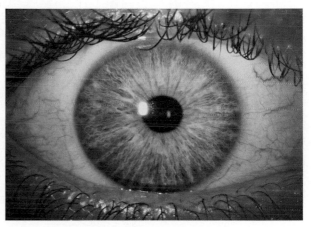

🔺 **Figure 8.7** The pupil dilates or constricts in response to changing light intensities.

Q8.3 Which of the eyes in Figure 8.7 is in dim light?

How do the muscles of the iris respond to light?

The iris controls the size of the pupil. It contains a pair of antagonistic muscles: radial and circular muscles (Figure 8.8). These are both controlled by the autonomic nervous system (see Topic 7, page 157). The radial muscles are like the spokes of a wheel, and are controlled by a sympathetic reflex. The circular muscles are controlled by a parasympathetic reflex. One reflex dilates the pupil; the other constricts it.

Q8.4 Which of the two sets of muscles will cause the pupil to dilate?

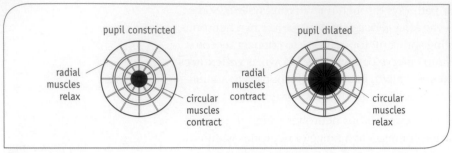

Figure 8.8 How the muscles act to constrict and dilate the pupil.

Controlling pupil size

High light levels striking the **photoreceptors** in the retina cause nerve impulses to pass along the optic nerve to a number of different sites within the CNS including a group of cells in the midbrain. Impulses from these coordinator cells are sent along parasympathetic motor neurones to the circular muscles of the iris, causing them to contract. At the same time the radial muscles relax. This constricts the pupil, reducing the amount of light entering the eye. Figure 8.9 shows the reflex pathway involved.

Figure 8.9 The reflex pathway involved in pupil constriction. Pupil dilation involves sympathetic neurones not shown here.

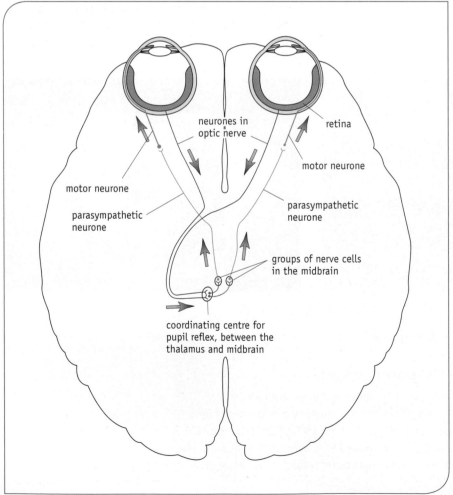

Activity

Investigate the pupil reflex in **Activity 8.1**. **A8.01S**

Q8.5 Name components a–d in Table 8.1 involved in the pupil reflex.

▼ **Table 8.1** Components involved in the pupil reflex.

Part of nervous system	Structure in the pupil reflex
receptors	a
sensory nerve fibres	b
coordinator	c
motor nerve fibres	oculomotor nerve
effector	d

Q8.6 What is the purpose of the pupil reflex?

Q8.7 The pupil reflex to increased light is very rapid. Why does this need to be the case?

Q8.8 How many synapses are there in the pupil reflex pathway shown in Figure 8.9?

Did you know? Atropine

Deadly nightshade (*Atropa belladonna*) is the source of the drug atropine. Atropine was used in the Middle Ages by women to make their pupils dilate. This was thought to make them more attractive to men, hence 'belladonna', Latin for 'beautiful lady', in the species name.

Atropine inhibits parasympathetic stimulation of the iris, so the circular muscles of the iris relax. Today the drug is used to dilate the pupils for an eye examination. It is known as an **acetylcholine** antagonist. When you have completed all the work on the nervous system you should understand what this means!

How nerve cells transmit impulses

To understand how Kenge saw, we must first understand how nerve cells transmit impulses, and how the receptor cells in the eye detect light, causing impulses to be sent to the brain where the signals are interpreted.

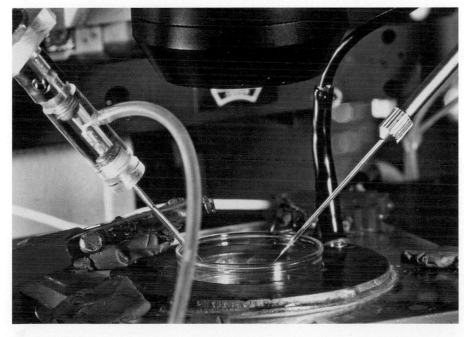

◀ **Figure 8.10** The giant axons of the squid can be seen by the naked eye, making it possible to manipulate them in experiments.

Much of the work done to establish what happens in a nerve fibre was carried out on the giant axons of the squid (Figure 8.10). The large size of these axons makes them easier to work with. Hodgkin, Huxley and Eccles carried out this work in the 1940s and 1950s, and they eventually won a Nobel Prize for their efforts.

Inside a resting axon

All cells have a potential difference (electrical voltage) across their surface membrane. Figure 8.11 shows an experimental setup designed to measure the potential difference across the membrane of an axon.

In Figure 8.11A, with both electrodes in the bathing solution, there is no potential difference. But if one of the electrodes is pushed inside the axon, as in Figure 8.11B, then the oscilloscope shows that there is a potential difference of around −70 millivolts. The inside of the axon is more negative than outside; the membrane is said to be polarised. The value of −70 mV is known as the **resting potential**.

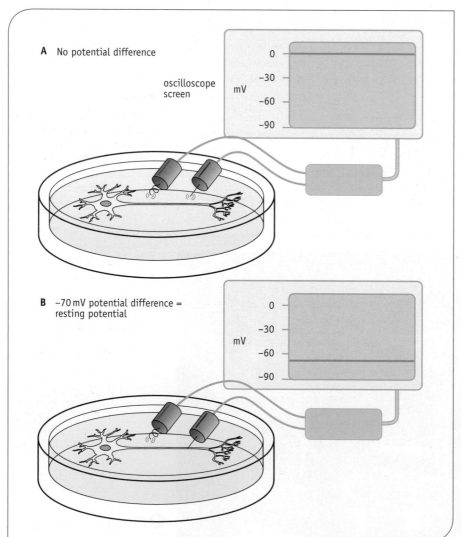

◀ **Figure 8.11** Measuring the potential difference across the axon membrane. The oscilloscope displays the potential difference between the two electrodes **A** when both electrodes are in the bathing solution and **B** across the axon membrane.

Key biological principle: Why is there a potential difference?

Sodium–potassium pumps

Table 8.2 shows the concentrations of some of the ions found in the solutions inside and outside a squid giant axon. The most obvious feature of this is that the distribution of the ions is far from equal.

Ion	Extracellular concentration	Intracellular concentration
K^+	30	400
Na^+	460	50
Cl^-	560	100
organic anions	0	370

Table 8.2 The approximate concentrations of ions inside and outside a nerve fibre ($mmol\ kg^{-1}$).

This uneven distribution of ions across the cell surface membrane is achieved by the action of **sodium–potassium pumps** in the cell surface membrane of the axon. These carry Na^+ out of the cell and K^+ into the cell (Figure 8.12). These pumps act against the concentration gradients of these two ions and are driven by energy supplied by hydrolysis of ATP. The organic anions (e.g. negatively charged amino acids) are large and stay within the cell, so chloride ions move out of the cell to balance the charge across the cell surface membrane.

The resting potential

Once these concentration gradients have been established by the sodium–potassium pumps and there is no difference in charge between the inside and outside of the membrane, potassium ions diffuse out of the neurone down the potassium concentration gradient. These K^+ pass through potassium channels, making the outside of the cell surface membrane positive and the inside negative. The membrane is permeable to potassium ions but is virtually impermeable to sodium ions. Although there is some leakage of Na^+ out of the neurone, down the Na^+ concentration gradient, it does not balance the difference in charge across the membrane. The difference in charge caused by diffusion of K^+ through the K^+ channels causes a potential difference across the membrane, and this is known as the resting potential (Figure 8.12).

Figure 8.12 Movement of sodium and potassium ions across the cell surface membrane of the neurone leads to the formation of the resting potential. At −70 mV, further movement of K^+ out of the cell due to the concentration gradient is opposed by the electrical gradient across the cell surface membrane.

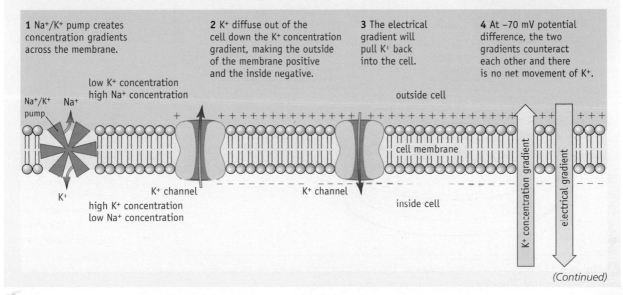

1 Na^+/K^+ pump creates concentration gradients across the membrane.

2 K^+ diffuse out of the cell down the K^+ concentration gradient, making the outside of the membrane positive and the inside negative.

3 The electrical gradient will pull K^+ back into the cell.

4 At −70 mV potential difference, the two gradients counteract each other and there is no net movement of K^+.

Na^+/K^+ pump

Na^+

low K^+ concentration
high Na^+ concentration

outside cell

cell membrane

K^+

K^+ channel

high K^+ concentration
low Na^+ concentration

K^+ channel

inside cell

K^+ concentration gradient

electrical gradient

(Continued)

Why is the axon resting potential –70 mV?

To understand why the axon resting potential is –70 mV we need to appreciate that there are two forces involved in the movement of the potassium ions. These result from:

- the concentration gradient generated by the Na^+/K^+ pump
- the electrical gradient due to the difference in charge on the two sides of the membrane resulting from K^+ diffusion.

Potassium ions diffuse out of the cell down the concentration gradient. The more potassium ions that diffuse out of the cell, the larger the potential difference across the membrane. The increased negative charge created inside the cell as a consequence attracts potassium ions back across the membrane into the cell. When the potential difference across the membrane is around –70 mV the electrical gradient exactly balances the chemical gradient. There is no net movement of K^+, and hence a steady state exists, maintaining the potential difference at –70 mV. An electrochemical equilibrium for potassium is in place and the membrane is polarised (Figure 8.12).

What happens when a nerve is stimulated?

Neurones are electrically excitable cells, which means that the potential difference across their cell surface membrane changes when they are conducting an impulse.

Figure 8.13 shows the effect of stimulating the axon with a small electric current. If an electric current above a threshold level is applied to the membrane, it causes a massive change in the potential difference. The potential difference across the membrane is locally reversed, making the inside of the axon positive and the outside negative. This is known as **depolarisation**.

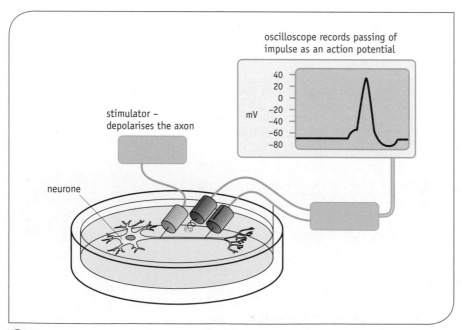

🔺 **Figure 8.13** Measuring an action potential. The stimulator produces an electric current causing the potential difference across the axon membrane to reverse.

The potential difference becomes +40 mV or so for a very brief instant, lasting about 3 milliseconds (ms), before returning to the resting state, as shown by the oscilloscope trace (Figure 8.13). It is important that the membrane is returned to the resting potential as soon as possible in order that more impulses can be conducted. This return to a resting potential of –70 mV is known as **repolarisation**. The large change in the voltage across the membrane is known as an **action potential**.

What causes an action potential?

Once threshold stimulation occurs, an action potential is caused by changes in the permeability of the cell surface membrane to Na^+ and K^+, due to the opening and closing of voltage-dependent Na^+ and K^+ channels (Figure 8.14). At the resting potential, these channels are blocked by gates preventing the flow of ions. Changes in the voltage across the membrane cause the gates to open, and so they are referred to as voltage-dependent channels. There are three stages in the generation of an action potential.

1 Depolarisation

When a neurone is stimulated some depolarisation occurs. The change in the potential difference across the membrane causes a change in the shape of the Na^+ gate, opening some of the voltage-dependent sodium ion channels. As the sodium ions flow in, depolarisation increases, triggering more gates to open once a certain threshold is reached. This increases depolarisation further. This is an example of **positive feedback** – a change encourages further change of the same sort, and it leads to a rapid opening of all the Na^+ gates. This means there is no way of controlling the degree of depolarisation of the membrane; action potentials are either there or they are not. This property is often referred to as **all-or-nothing**

There is a higher concentration of sodium ions outside the axon, so sodium ions flow rapidly inwards through the open voltage-dependent Na^+ channels, causing a build-up of positive charges inside. This reverses the polarity of the membrane. The potential difference across the membrane reaches +40 mV.

2 Repolarisation

After about 0.5 ms, the voltage-dependent Na^+ channels spontaneously close and Na^+ permeability of the membrane returns to its usual very low level. Voltage-dependent K^+ channels open due to the depolarisation of the membrane. As a result, potassium ions move out of the axon, down the electrochemical gradient (they diffuse down the concentration gradient and are also attracted by the negative charge outside the cell surface membrane). As potassium ions flow out of the cell, the inside of the cell once again becomes more negative than the outside. This is the falling phase of the oscilloscope trace in Figure 8.14.

3 Restoring the resting potential

The membrane is highly permeable to potassium, and more ions move out than occurs at the resting potential, making the potential difference more negative (Figure 8.14). This is known as **hyperpolarisation** of the membrane. The resting potential is re-established by closing of the voltage-dependent K^+ channels and potassium ions diffusing into the axon.

Activity

Use interactive tutorial **Activity 8.2** to investigate an action potential in detail. **A8.02S**

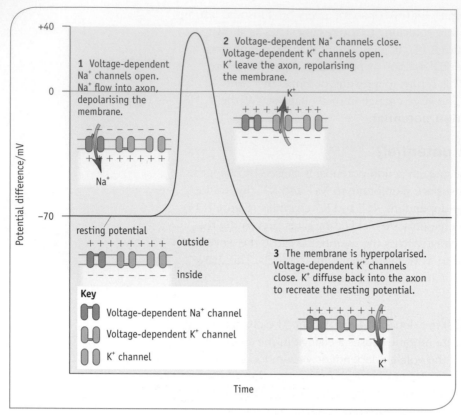

◀ **Figure 8.14** Voltage-dependent gates open and close to produce changes in potential difference during an action potential.

If lots (hundreds) of action potentials occur in the neurone, the sodium ion concentration inside the cell rises significantly. The sodium–potassium pumps start to function, restoring the original ion concentrations across the cell membrane (Table 8.2). If a cell is not transmitting many action potentials, these pumps will not have to be used very frequently. At rest there is some slow leakage of sodium ions into the axon. These sodium ions are pumped back out of the cell.

Q8.9 Why is the Na⁺ channel described above referred to as a voltage-dependent channel?

Q8.10 Will it be possible for an action potential to be triggered in a dead axon? Give a reason for your answer.

How is the impulse passed along an axon?

When a neurone is stimulated, the action potential generated does not actually travel along the axon, but triggers a sequence of action potentials along the length of the axon. It is rather like pushing one domino to topple a whole line of standing dominoes. Figure 8.15 illustrates this propagation of the impulse along an axon.

As part of the membrane becomes depolarised at the site of an action potential, a local electric current is created as the electrically charged sodium ions flow between the depolarised part of the membrane and the adjacent resting region. The depolarisation spreads to the adjacent region and the nearby Na⁺ gates will respond to this by opening as described earlier, triggering another action potential. These events are then repeated along the membrane. As a result, a wave of depolarisation will pass along the membrane. This is the nerve impulse.

Checkpoint

8.2 Produce a bullet-point summary of the membrane changes and ion movements that cause an action potential. You should aim to have at least 10 bullet points.

Activity

Use **Activity 8.3** to help understand how the nerve impulse is transmitted along the axon. **A8.03S**

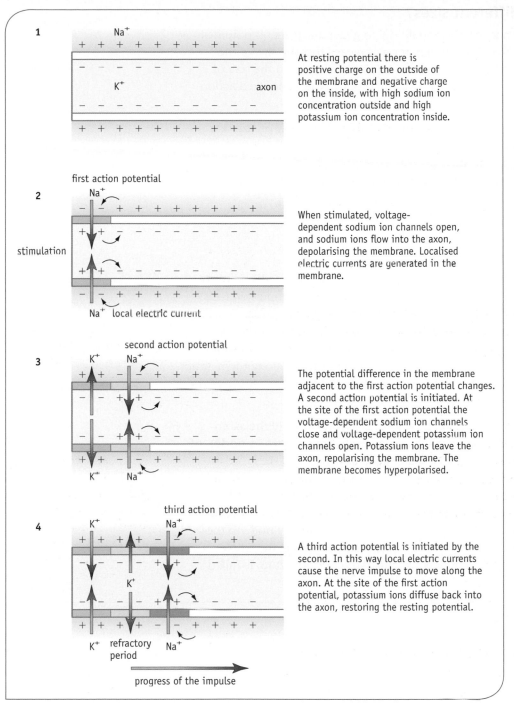

Figure 8.15 Propagation of an impulse along an axon.

A new action potential cannot be generated in the same section of membrane for a millisecond or two. This is known as the **refractory period**. It lasts until all the voltage-dependent sodium and potassium channels have returned to their normal resting state (closed) and the resting potential is restored. The refractory period ensures that impulses travel in only one direction.

Q8.11 How will the refractory period ensure that an action potential will not be propagated back the way it came?

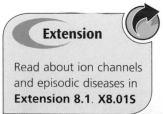

Extension

Read about ion channels and episodic diseases in **Extension 8.1. X8.01S**

Are impulses different sizes?

A stimulus must be above a threshold level to generate an action potential. The all-or-nothing effect for action potentials means that the size of the stimulus, assuming it is above the threshold, has no effect on the size of the action potential. A very strong light will produce the same sized action potential in a neurone from your eye as does a dim light.

Different mechanisms are used to communicate the intensity or size of a stimulus. The size of stimulus affects:

- the frequency of impulses
- the number of neurones in a nerve that are conducting impulses.

A high frequency of firing and the firing of many neurones are usually associated with a strong stimulus.

Speed of conduction

The speed of nervous conduction is in part determined by the diameter of the axon. In general, the wider the diameter the faster the impulse travels. The normal axons of a squid (diameter 1–20 μm) conduct impulses at around 0.5 m s^{-1}, whereas the giant axons (diameter up to 1000 μm) conduct at nearer 100 m s^{-1}. The nerve axons of mammals are much finer than squid giant axons, about 1–20 μm in diameter, but impulses travel along them at up to 120 m s^{-1}. This apparent anomaly can be explained by the presence of the myelin sheath around mammalian nerve axons.

The myelin sheath acts as an electrical insulator along most of the axon, preventing any flow of ions across the membrane. Gaps known as **nodes of Ranvier** occur in the myelin sheath at regular intervals, and these are the only places where depolarisation can occur. As ions flow across the membrane at one node during depolarisation, a circuit is set up which reduces the potential difference of the membrane at the next node, triggering an action potential. In this way, the impulse effectively jumps from one node to the next. This is much faster than a wave of depolarisation along the whole membrane. This 'jumping' conduction, illustrated in Figure 8.16, is called **saltatory conduction**.

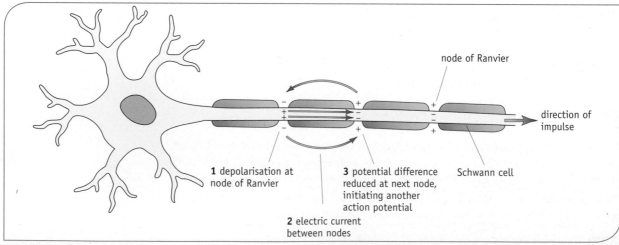

▲ **Figure 8.16** An impulse can move very quickly along the axon by jumping between the nodes of Ranvier.

How does an impulse pass between cells?

Where two neurones meet is known as a **synapse**. The cells do not actually touch – there is a small gap, the **synaptic cleft**. So how does the nerve impulse, on which the function of the nervous system depends, get across this gap?

Synapse structure

A nerve cell may have very large numbers of synapses with other cells (Figure 8.3), possibly as many as 10000 for some brain cells. This is important in enabling the distribution and processing of information.

Figure 8.17 shows the structure of a typical synapse. Notice the synaptic cleft that separates the **presynaptic** membrane of the stimulating neurone from the **postsynaptic** membrane of the other cell. The gap is about 20–50 nm and a nerve impulse cannot jump across it. In the cytoplasm at the end of the presynaptic neurone there are numerous **synaptic vesicles** containing a chemical called a **neurotransmitter**.

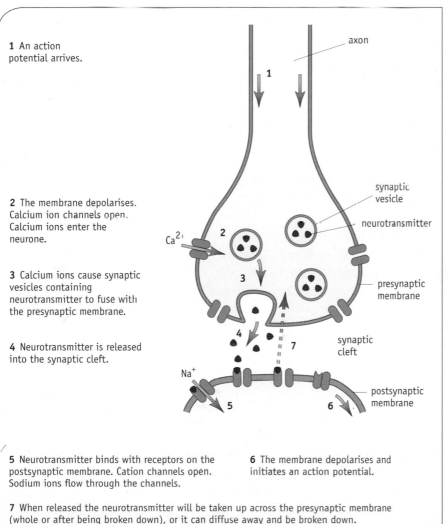

1 An action potential arrives.

2 The membrane depolarises. Calcium ion channels open. Calcium ions enter the neurone.

3 Calcium ions cause synaptic vesicles containing neurotransmitter to fuse with the presynaptic membrane.

4 Neurotransmitter is released into the synaptic cleft.

axon

synaptic vesicle

neurotransmitter

presynaptic membrane

synaptic cleft

postsynaptic membrane

5 Neurotransmitter binds with receptors on the postsynaptic membrane. Cation channels open. Sodium ions flow through the channels.

6 The membrane depolarises and initiates an action potential.

7 When released the neurotransmitter will be taken up across the presynaptic membrane (whole or after being broken down), or it can diffuse away and be broken down.

◀ **Figure 8.17** The functioning of a synapse.

How does the synapse transmit an impulse?

The arrival of an action potential at the presynaptic membrane causes the release of the neurotransmitter into the synaptic cleft. The neurotransmitter diffuses across the gap, resulting in events that cause the depolarisation of the postsynaptic membrane, and hence the propagation of the impulse along the next cell. The presynaptic cell expends a considerable amount of energy to produce neurotransmitter and to package it into vesicles ready for transport out of the cell.

Many neurotransmitters have been discovered, with 50 identified in the human central nervous system. **Acetylcholine**, the first to be discovered, will be used here to describe the working of a synapse. Others will be considered later in the topic.

There are essentially three stages leading to the nerve impulse passing along the postsynaptic neurone:
- neurotransmitter release
- stimulation of the postsynaptic membrane
- inactivation of the neurotransmitter.

These stages are illustrated in Figure 8.17.

Neurotransmitter release

When the presynaptic membrane is depolarised by an action potential, channels in the membrane open and increase the permeability of the membrane to calcium ions (Ca^{2+}). These calcium ions are in greater concentration outside the cell, so they diffuse across the membrane and into the cytoplasm.

The increased Ca^{2+} concentration causes synaptic vesicles to fuse with the presynaptic membrane and release their contents into the synaptic cleft by exocytosis.

Stimulation of the postsynaptic membrane

The neurotransmitter takes about 0.5 ms to diffuse across the synaptic cleft and reach the postsynaptic membrane. Embedded in the postsynaptic membrane are specific receptor proteins that have a binding site with a complementary shape to part of the acetylcholine molecule. The acetylcholine molecule binds to the receptor, changing the shape of the receptor protein, opening cation channels and making the membrane permeable to sodium ions. The flow of sodium ions across the postsynaptic membrane causes depolarisation, and if there is sufficient depolarisation, an action potential will be produced and propagated along the postsynaptic neurone.

The extent of the depolarisation will depend on the amount of acetylcholine reaching the postsynaptic membrane. This will depend in part on the frequency of impulses reaching the presynaptic membrane. A single impulse will not usually be enough; several impulses are usually required to generate enough neurotransmitter to depolarise the postsynaptic membrane. The number of functioning receptors in the postsynaptic membrane will also influence the degree of depolarisation.

Inactivation of the neurotransmitter

Some neurotransmitters are actively taken up by the presynaptic membrane and the molecules are used again. In others, the neurotransmitter rapidly diffuses away from the synaptic cleft or is taken up by other cells of the nervous system. In the case of acetylcholine, a specific enzyme, **acetylcholinesterase**, at the postsynaptic membrane breaks down the acetylcholine so that it can no longer bind to receptors. Some of the breakdown products are then reabsorbed by the presynaptic membrane and reused.

What is the role of synapses in nerve pathways?

Control and coordination

Synapses have two roles:

- control of nerve pathways allowing flexibility of response
- integration of information from different neurones allowing a coordinated response.

The postsynaptic cell is likely to be receiving input from many synapses at the same time (Figure 8.18). It is the overall effect of all of these synapses that will determine whether the postsynaptic cell generates an action potential. Two main factors affect the likelihood that the postsynaptic membrane will depolarise:

- the type of synapse
- the number of impulses received.

Some synapses help stimulate an action potential, whereas others are inhibitory. They make it less likely that the postsynaptic membrane will depolarise. A postsynaptic cell can have many inhibitory and excitatory synapses, and so whether or not an action potential results depends upon the balance of excitatory and inhibitory synapses acting at any given time.

Activity

You can investigate the synapse in more detail using the animation in **Activity 8.4. A8.04S**

Checkpoint

8.3 Construct a flowchart to show the sequence of events that occurs when a nerve impulse crosses a synapse.

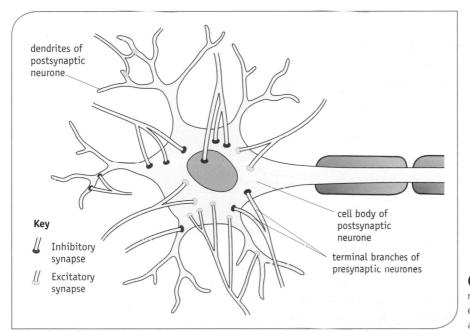

Key

- Inhibitory synapse
- Excitatory synapse

dendrites of postsynaptic neurone

cell body of postsynaptic neurone

terminal branches of presynaptic neurones

Figure 8.18 A postsynaptic neurone receives input from many excitatory and inhibitory synapses at the same time.

Types of synapse

Excitatory synapses

Excitatory synapses make the postsynaptic membrane more permeable to sodium ions. A single excitatory synapse typically does not depolarise the membrane enough to produce an action potential, but several impulses added together produce sufficient depolarisation via the release of neurotransmitter to produce an action potential in the postsynaptic cell. The fact that each impulse adds to the effect of the others is known as **summation**.

There are two types of summation:

- **Spatial summation** – here the impulses are from different synapses, usually from different neurones. The number of different sensory cells that have been stimulated can be reflected in the control of the response (Figure 8.19A).
- **Temporal summation** – in this case several impulses arrive at a synapse having travelled along a single neurone one after the other, each causing the release of neurotransmitter. Their combined effect may be to generate an action potential in the postsynaptic membrane (Figure 8.19B).

Generally, the more intense the stimulus detected by a sensory cell, the more impulses it will generate. In this way more intense stimuli are more likely to cause a response. One small insect (e.g. a thrip) landing on your arm may not be noticed, but a larger one (e.g. a butterfly) crawling along your arm would stimulate lots of receptors over a longer period of time, causing you to respond.

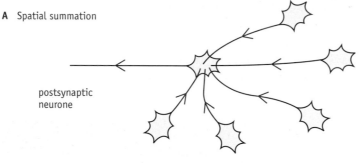

A Spatial summation

postsynaptic
neurone

Impulses from several different neurones produce an action
potential in the postsynaptic neurone.

B Temporal summation

postsynaptic
neurone

Several impulses along one neurone produce an action
potential in the postsynaptic neurone.

▲ **Figure 8.19** Spatial and temporal summation.

Inhibitory synapses

Inhibitory synapses make it less likely that an action potential will result in the postsynaptic cell. The neurotransmitter from these synapses opens channels for chloride ions and potassium ions in the postsynaptic membrane, and these ions will then follow their diffusion gradients (Figure 8.20). Chloride ions will move into the cell, carrying in negative charge, and potassium ions will move out, carrying out positive charge. The result will be a *greater* potential difference across the membrane as the inside will become more negative than usual (about −90 mV), so-called hyperpolarisation. This makes depolarisation less likely, as more excitatory synapses are required to depolarise the membrane.

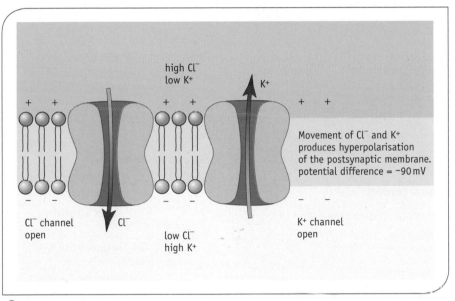

▲ **Figure 8.20** Neurotransmitters from inhibitory synapses open Cl⁻ and K⁺ channels in the postsynaptic membrane.

Key biological principle: Comparing nervous and hormonal coordination

Hormonal coordination
The nervous system is not the only means by which the activities of the body can be coordinated. Chemicals called **hormones** act as a means of chemical communication with target cells. They are secreted by **endocrine glands** (Figure 8.21 and Table 8.3) into the bloodstream and are transported throughout the body.

Each hormone affects only specific target cells, modifying their activity. Some hormones bind to receptors on the cell surface, producing a second messenger that can activate enzymes within the cell; others act on the cell by direct or indirect signalling to control transcription of enzymes (see Topic 3, signal proteins).

Many hormones are produced steadily over long periods of time to control long-term changes in the body, such as growth and sexual development. See, for instance, the effects of testosterone in Topic 7. Adrenaline, also covered in Topic 7, is more short term in its action, but is slow in producing a response compared with the action of the nervous system.

(Continued)

Gland	Hormone	Function
pituitary gland	growth hormone follicle-stimulating hormone antidiuretic hormone	stimulates growth controls testes and ovaries causes reabsorption of water in kidneys
thyroid gland	thyroxine	raises basal metabolic rate
adrenal gland	adrenaline	raises basal metabolic rate, dilates blood vessels, prepares the body for action
pancreas	insulin	lowers blood glucose concentration
ovary	oestrogen	promotes development of ovaries and female secondary sexual characteristics
testis	testosterone	promotes development of male secondary sexual characteristics

heart — stomach — kidney

△ **Figure 8.21** and **Table 8.3** The main endocrine glands, with some examples of the hormones they produce.

Comparing nervous and hormonal control

Q8.12 Look back at Topic 7 and describe in detail the action of the hormone testosterone. Make sure you include:
- site of production
- method of transport
- location of target cells
- effect on the target cells.

Table 8.4 contrasts nervous and hormonal control in animals.

▼ **Table 8.4** Nervous and hormonal control in animals.

Nervous control	Hormonal control
electrical transmission by nerve impulses and chemical transmission at synapses	chemical transmission; carried in the blood
rapid acting	slower acting
usually associated with short-term changes, e.g. muscle contraction	can control long-term changes, e.g. growth
action potentials carried by neurones with connections to specific cells	blood carries the hormone to all cells, but only target cells are able to respond
response is often very local, such as by a specific muscle cell or gland	response may be widespread, such as in growth and development

Checkpoint

8.4 Make a list of key words to distinguish the features of: **a** nervous and **b** hormonal control. If you can compose a mnemonic to help you remember these, even better.

Extension

In **Extension 8.2** read about how scientists link sex hormones with brain activities. **X8.02S**

8.2 Reception of stimuli

How does light trigger nerve impulses?

When reflected light entered Kenge's eye, how was the light converted into electrical impulses which could be passed along the optic nerve to the brain?

Receptors

Stimuli (any changes that occur in an animal's environment) are detected by receptor cells that send electrical impulses to the central nervous system. Many receptors are spread throughout the body, but some types of receptor cell are grouped together into **sense organs**. Sense organs such as eyes help to protect the receptor cells and improve their efficiency; structures within the sense organ ensure that the receptor cells are able to receive the appropriate stimulus. The receptor cells that detect light are found in the eye. The lens and cornea refract (bend) the light so that it is focused on the retina where the photoreceptor cells are located.

Activity

Remind yourself about the structure and function of the different parts of the eye in **Activity 8.5**. **A8.05S**

> **Did you know?** Different types of receptor
>
> Receptors allow us to perceive and respond to a wide variety of stimuli. The receptors can either be cells that synapse with a sensory neurone, or can be part of a specialised sensory neurone, like the temperature receptors in the skin (see Figure 6.45). Four of the main types of receptor are shown in Table 8.5.
>
> ▼ **Table 8.5** Four types of receptor and their roles.
>
> | Type of receptor | Stimulated by | Examples of role in body |
> |---|---|---|
> | mechanoreceptors | forces that stretch, compress or move the sensor | balance, touch and hearing |
> | chemoreceptors | chemicals | taste, smell and regulation of chemical concentrations in the blood |
> | thermoreceptors | temperature | thermoregulation and awareness of changes in the surrounding temperature |
> | photoreceptors | light | sight |
>
> All these receptors, except photoreceptors, work in a similar manner. At rest, the cell surface membrane has a negative resting potential. Stimulation of the receptor causes depolarisation of the cell. The stronger the stimulus, the greater the depolarisation. When depolarisation exceeds the threshold level, it triggers an action potential. This is either relayed across the synapse using neurotransmitters or passed directly down the axon of the sensory nerve.

Before addressing in detail the question 'How does light trigger nerve impulses?' remember what you have learned before about the way that the parts of the eye work together, using Figure 8.22 and the revision quiz in Activity 8.5.

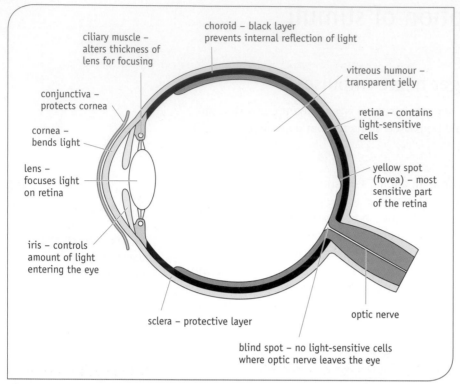

ciliary muscle –
alters thickness of
lens for focusing

choroid – black layer
prevents internal reflection of light

vitreous humour –
transparent jelly

conjunctiva –
protects cornea

retina – contains
light-sensitive
cells

cornea –
bends light

yellow spot
(fovea) – most
sensitive part
of the retina

lens –
focuses light
on retina

iris – controls
amount of light
entering the eye

sclera – protective layer

optic nerve

blind spot – no light-sensitive cells
where optic nerve leaves the eye

⚠ Figure 8.22 The structure of the eye.

Photoreceptors

The human retina contains two types of photoreceptor cell sensitive to light: **rods** and **cones** (see Figure 8.23). Cones allow colour vision in bright light; rods give only black-and-white vision but, unlike cones, work in dim light conditions. In the centre of the retina, in an area about the size of this 'o', there are only cones. This area allows us to pinpoint accurately the source and detail of what we are looking at. Over the remainder of the retina, the rods outnumber the cones in a ratio of about 20:1.

In Figure 8.23 notice the arrangement of the three layers of cells that make up the retina; light hitting the retina has to pass through the layers of neurones *before* reaching the rods and cones. The rods and cones synapse with **bipolar neurone** cells, which in turn synapse with **ganglion neurones**, whose axons together make up the **optic nerve**.

Q8.13 Can you explain why some people describe the retina as *functionally inside out*?

How does light stimulate photoreceptor cells?

In both rods and cones, a photochemical pigment absorbs light, resulting in a chemical change. In rods this is a reddish pigment called **rhodopsin**. In Figure 8.24 you can see the rod cell outer and inner segments; these contain many layers of flattened vesicles. The rhodopsin molecules are located in the membranes of these vesicles.

Extension

Read **Extension 8.3** to find out how cones function to allow colour vision. **X8.03S**

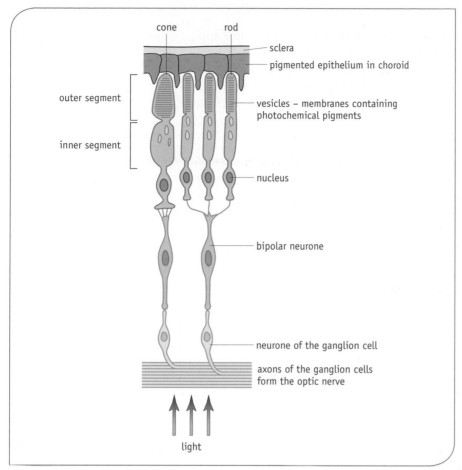

Figure 8.23 The structure of rods and cones within the retina.

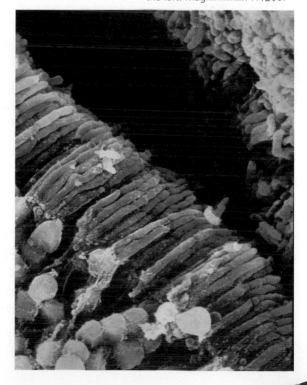

Figure 8.24 An electron micrograph of rods shows the outer segments that contain the photochemical pigment. At the top right the choroid layer, normally attached to the retina, has peeled away. The rods synapse with the neurones on the left. Magnification ×1200.

Look at the diagram of the rod cell in Figure 8.23 and notice that it has an inner and outer segment. Figure 8.25 summarises the processes that occur in the light and dark within the rod cells.

In the dark

In the dark, sodium ions flow into the outer segment through **non-specific cation channels**. The sodium ions move down the concentration gradient into the inner segment where pumps continuously transport them back out of the cell. The influx of Na^+ ions produces a slight depolarisation of the cell. The potential difference across the membrane is about −40 mV, compared to the −70 mV resting potential of a cell. This slight depolarisation triggers the continual release of a neurotransmitter, thought to be **glutamate**, from the rod cells. In the dark, rods release this neurotransmitter continuously. The neurotransmitter binds to the bipolar neurone, stopping it depolarising.

In the light

When light falls on the rhodopsin molecule, it breaks down into retinal and opsin. The opsin activates a series of membrane-bound reactions, ending in hydrolysis of a molecule attached to the cation channel in the outer segment. The hydrolysis results in the closing of the cation channels. The influx of Na^+ into the rod decreases, while the inner segment continues to pump Na^+ out. This makes the inside of the cell more negative. It becomes hyperpolarised, and the release of the neurotransmitter glutamate stops.

The lack of glutamate results in depolarisation of the bipolar cell with which the rod synapses. Cation channels in the bipolar cell membrane open; these channels are normally prevented from opening by the neurotransmitter. The neurones that make up the optic nerve are also depolarised and respond by producing an action potential.

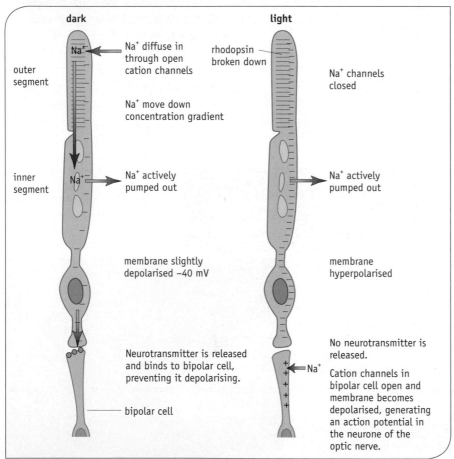

△ **Figure 8.25** A rod cell in the dark and in the light.

Q8.14 **a** By what form of transport will sodium ions: **i** be pumped out of the rod cell **ii** flow back into the cell?
b What do you think a 'non-specific cation channel' is?
c Why does the rod cell membrane become hyperpolarised in the light?

Checkpoint ✓

8.5 Produce a series of statements that describe what happens in rod cells to enable light to generate an action potential in a bipolar cell. Get a friend to order the statements as a revision exercise.

Once the rhodopsin has been broken down, it is essential for it to be rapidly converted back to its original form so that repeated stimuli can be perceived. Each individual rhodopsin molecule takes a few minutes to be reformed. The higher the light intensity, the more rhodopsin molecules are broken down and the longer it can take for all the rhodopsin to reform, up to a maximum of 50 minutes. This reforming of rhodopsin is called **dark adaptation**.

Activity

Try **Activity 8.6** to experience the effects of dark adaptation, and think through how the rod cells work. **A8.06S**

Did you know? Processing starts in the retina

Processing of the signals (nerve impulses) from photoreceptors starts within the retina.

- Bipolar cells may receive signals not only from photoreceptors but also from adjacent bipolar cells.
- If photoreceptors are stimulated they inhibit their neighbours. Higher intensity light has a greater inhibiting effect. This is called lateral inhibition.

Information from the retina does not inform the brain exactly what the light intensity is at every point. Instead, it highlights where there are changes in the intensity. So the perceived brightness of two adjoining areas is not a reflection of the actual light intensity stimulating the cells in that part of the retina, but an impression of the relative light intensities. This visual contrast can be demonstrated using a simple set of shaded squares, as shown in Figure 8.26.

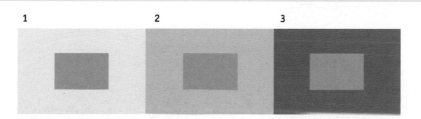

△ **Figure 8.26** All the central rectangles are the same shade and reflect the same amount of light onto the retina, but due to lateral inhibition they are seen as different shades. The photoreceptors stimulated by light reflected from the area surrounding the central rectangle in 1 receive a greater light intensity. Therefore they inhibit the photoreceptors detecting the central rectangle. This makes the central rectangle appear darker. In 3 the central rectangle reflects more light than the surrounding area; therefore the photoreceptors detecting the central rectangle inhibit the photoreceptors detecting the outer region. The central rectangle appears lighter.

Did you know? Why you should eat your carrots

Have you ever been told to finish your carrots so that you will be able to see in the dark? And did you believe it? Like many old wives' tales there is a grain of truth in this one.

Poor night vision, sometimes called night blindness, has been known for many years to be one of the symptoms of the disease caused by a shortage of vitamin A in the diet.

Vitamin A is closely related to a substance called retinal, which is part of the rhodopsin found in the rods. A shortage of vitamin A leads to a lack of retinal and thus rhodopsin, which means poor vision in low light conditions.

From the eye to the brain

The axons of the ganglion cells that make up the optic nerve pass out of the eye and extend to several areas of the brain, including a part of the thalamus as shown in Figure 8.27. The impulses are then sent along further neurones to the primary visual cortex where the information is processed further.

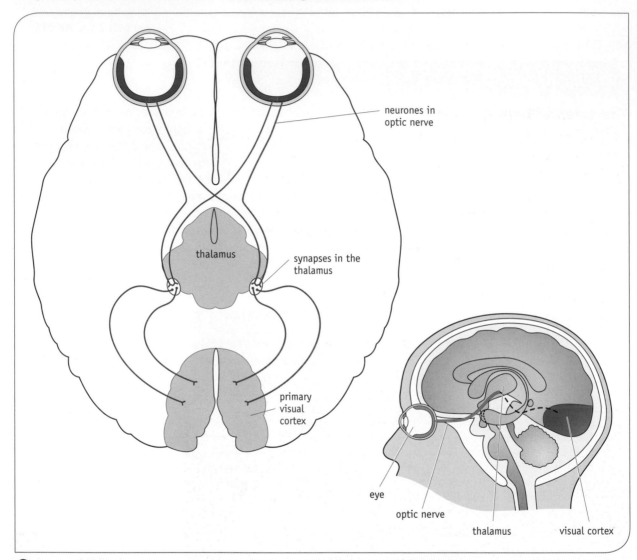

neurones in optic nerve

thalamus

synapses in the thalamus

primary visual cortex

eye

optic nerve

thalamus

visual cortex

🔺 **Figure 8.27** The visual pathway. The right side of the brain interprets input from the right side of the retina of both eyes, that is, the left side of the field of view. The left side of the brain interprets information from the left side of both retinas.

Before reaching the thalamus, some of the neurones in each optic nerve branch off to the midbrain, where they connect to motor neurones involved in controlling the pupil reflex (Figure 8.9) and movement of the eye. Audio signals also arrive at the midbrain so we can quickly turn our eyes in the direction of a visual or auditory stimulus.

To discover where the visual cortex is found in the brain and how it develops, read Sections 8.3 (page 207) and 8.4 (page 214). You can then find out what happens in the visual cortex in Section 8.5 (page 219).

8.3 The brain

Regions of the brain

The brain is the main coordinating centre for nervous activity, receiving information from sense organs, interpreting it, and then transmitting information to effectors. Different regions within the brain are involved in helping us respond to our external environment and regulating our internal environment (see Figures 8.29 and 8.30).

The cerebral hemispheres

Looking down at the brain from the top (Figure 8.28), you see the **cortex**. It is grey and highly folded, composed mainly of nerve cell bodies, synapses and dendrites. This thin outer layer of the brain is known as the **grey matter**.

The cortex, accounting for about two-thirds of the human brain's mass, is the largest region of the brain. It is positioned over and around most other brain regions, and is divided into the left and right **cerebral hemispheres**. Each hemisphere is composed of four regions called lobes (Figure 8.29):

- **frontal lobe**
- **parietal lobe**
- **occipital lobe**
- **temporal lobe**.

Each lobe interprets and manages its own separate sensory inputs. The two cerebral hemispheres are connected by a sweeping band of **white matter** (nerve axons) called the corpus callosum.

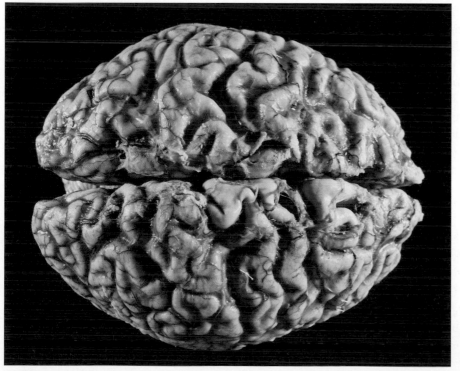

Activity

Try **Activity 8.7** to gain an impression of the size of the cortex. **A8.07S**

◀ **Figure 8.28** The two cerebral hemispheres can easily be seen from above.

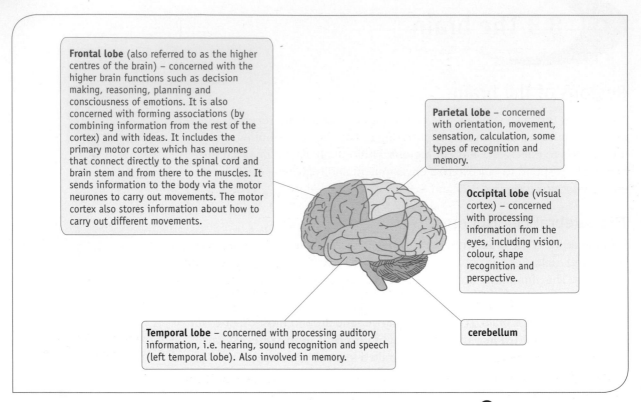

Frontal lobe (also referred to as the higher centres of the brain) – concerned with the higher brain functions such as decision making, reasoning, planning and consciousness of emotions. It is also concerned with forming associations (by combining information from the rest of the cortex) and with ideas. It includes the primary motor cortex which has neurones that connect directly to the spinal cord and brain stem and from there to the muscles. It sends information to the body via the motor neurones to carry out movements. The motor cortex also stores information about how to carry out different movements.

Parietal lobe – concerned with orientation, movement, sensation, calculation, some types of recognition and memory.

Occipital lobe (visual cortex) – concerned with processing information from the eyes, including vision, colour, shape recognition and perspective.

Temporal lobe – concerned with processing auditory information, i.e. hearing, sound recognition and speech (left temporal lobe). Also involved in memory.

cerebellum

▲ **Figure 8.29** The regions of the cerebral hemispheres and their functions. The location of the cerebellum is also shown.

The **basal ganglia** are a collection of neurones that lie deep within each hemisphere and are responsible for selecting and initiating stored programmes for movement.

Q8.15 A blow to the back of your head may result in your seeing stars. Suggest why.

The structures lying directly below the corpus callosum include, among others, the thalamus, the hypothalamus and the hippocampus.
- The **thalamus** is responsible for routing all the incoming sensory information to the correct part of the brain, via the axons of the white matter.
- The **hypothalamus** lies below the thalamus and contains the thermoregulatory centre. This monitors core body temperature and skin temperature, and initiates corrective action to restore the body to its optimum temperature (see Topics 6 and 7). Also located in the hypothalamus are other centres that control sleep, thirst and hunger. The hypothalamus acts as an endocrine gland, secreting hormones such as antidiuretic hormone (this controls water reabsorption in the kidneys and hence controls blood concentration). The hypothalamus connects directly to the **pituitary gland** which, in turn, secretes other hormones (see Figure 8.21, page 200).
- The **hippocampus** is involved in laying down long-term memory.

The cerebellum and brain stem

The brain stem is, in evolutionary terms, the oldest part of the brain and is sometimes referred to as the reptilian brain. It lies at the top of the spinal column. The brain stem extends from the midbrain to the **medulla oblongata** (Figure 8.30).

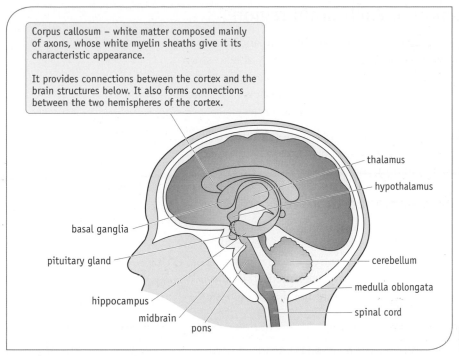

Corpus callosum – white matter composed mainly of axons, whose white myelin sheaths give it its characteristic appearance.

It provides connections between the cortex and the brain structures below. It also forms connections between the two hemispheres of the cortex.

thalamus

hypothalamus

basal ganglia

pituitary gland

cerebellum

medulla oblongata

hippocampus

spinal cord

midbrain

pons

◀ **Figure 8.30** Structures beneath the corpus callosum.

Study Table 8.6 which provides information about the role of the **cerebellum** and regions in the brain stem, and then try the questions which follow.

▼ **Table 8.6** The cerebellum and brain stem functions.

Brain region	Functions
cerebellum	• responsible for balance • coordinates movement as it is carried out, receiving information from the primary motor cortex, muscles and joints • constantly checks whether the motor programme being used is the correct one, for example by referring to incoming information about posture and external circumstances
midbrain	relays information to the cerebral hemispheres, including auditory information to the temporal lobe and visual information to the occipital lobe
medulla oblongata	regulates those body processes that we do not consciously control, such as heart rate, breathing and blood pressure

Q8.16 Imagine that you are whizzing downhill on a bike (Figure 8.31) and come across an unexpected sharp bend in the road. You need to apply the brakes or turn the handlebars to stop yourself falling off. Which regions of the brain, including those in the brain stem, are involved in your subsequent action?

Q8.17 Various diseases or conditions can give us insights into the functioning of the different areas of the brain. Research has shown that in Parkinson's disease neurones in a particular area of the brain have died. Parkinson's disease results in an inability to select and make appropriate movements. Suggest which lobe of the brain is damaged.

▶ **Figure 8.31** Aargh!

Discovering the function of each brain region

How do neuroscientists identify the different regions of the brain and know what each does? Until relatively recently, neuroscientists were only able to study the brain by looking at pathological specimens, examining the effect of damage to particular areas of the brain, through studies using animal models (see page 216) and, to some degree, by studying human patients during surgery.

Individuals with brain damage still provide valuable information in the study of the brain. However, neuroscientists and neurologists now have a wide range of non-invasive imaging techniques for studying the function of the living brain.

Studies of individuals with damaged brain regions

By studying the consequences of accidental brain damage it is possible to determine the functions of certain regions of the brain, as the examples that follow illustrate. Researchers have also studied the consequences of injuring or destroying neurones to produce lesions (areas of tissue destruction) in non-human animal 'models', and the consequences of the removal of brain tissue.

The story of Phineas Gage

Phineas Gage was the foreman of a railway construction company; a hard-working, fit, popular and responsible man. One day in 1848 he was working with dynamite when an explosion propelled a three-and-a-half-foot long iron bar through his head (Figure 8.32). Amazingly Gage didn't die, and although most of the front part of the left-hand side of his brain was destroyed, he could still walk and talk.

But after the accident Gage's personality changed: he became nasty, foul-mouthed and irresponsible. He was also impatient and obstinate and was unable to complete any plans for future action.

Phineas Gage died 12 years later. Researchers at the Harvard University Medical School have since combined photographs and X-rays of Gage's skull with computer graphics to determine the areas of his brain that would have been damaged by the bar. It is highly probable that the accident severed connections between his midbrain and frontal lobes. Gage's reduced ability to control his emotional behaviour after the accident was related to damage at this site.

The strange case of Lincoln Holmes

Imagine what it would be like to never be able to put a name to a face. That's what life is like for Lincoln Holmes. He finds recognising a face impossible. Thirty years ago a car accident left him with damage to an isolated part of his temporal lobe and he is now 'face-blind'. Even when shown a photograph of himself, he has to be prompted before he realises he is staring at his own image. Lincoln can see facial features, but they all appear as a jumble and he is unable to put all the component parts together. Lincoln's case has revealed that recognition of faces is at least partly carried out by a specific face-recognition unit, in the temporal lobe.

Q8.18 Physical damage is one obvious cause of brain damage. Can you think of any others?

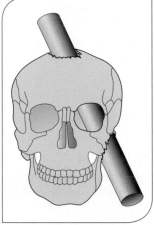

▲ **Figure 8.32** The iron bar travelled behind Gage's left eye and flew out through the top of his skull.

Weblink

Listen to Lincoln Holmes on the BBC News website. Damage to an area of his temporal lobe left him 'face-blind'.

Activity

In **Activity 8.8** you can identify regions of the brain by considering symptoms that occur after damage to that area. **A8.08S**

The effects of strokes

Brain damage caused by a stroke (Figure 8.33) can cause an impaired ability to speak, trouble understanding speech, and difficulty with reading and writing. In the nineteenth century Paul Broca studied several post-mortems of patients who could not speak due to strokes. He concluded that lesions in a small cortical area in the left frontal lobe (subsequently known as Broca's area) were responsible for deficits in language production.

Some patients can recover some abilities after a stroke, showing the potential of neurones to change in structure and function. This is known as **neural plasticity**. The structure of the brain remains flexible even in later life and can respond to changes in the environment. Brain structure and functioning is affected by both nature and nurture.

Brain imaging

CT scans

Computerised axial tomography (**CT** or **CAT**) imaging was developed in the 1970s to overcome the limitations of X-rays. Standard broad-beam X-rays cannot be used for imaging soft tissue as they are only absorbed by denser materials such as bone.

CT scans use thousands of narrow-beam X-rays rotated around the patient to pass through the tissue from different angles. Each narrow beam is attenuated (reduced in strength) according to the density of the tissue in its path. The X-rays are detected and are used to produce an image of a thin slice of the brain on a computer screen in which the different soft tissues within the brain can be distinguished (Figure 8.33).

CT scans give only 'frozen moment' pictures. They look at structures in the brain rather than at functions, and are used to detect brain disease and to monitor the tissues of the brain over the course of an illness. However, they have relatively low resolution so small structures in the brain cannot be distinguished.

Techniques that do not rely on harmful X-rays and can therefore be used more frequently have been developed, including magnetic resonance imaging.

Magnetic resonance imaging

In Topic 1 we saw how **magnetic resonance imaging** (**MRI**) scans helped to diagnose Mark's stroke. Here we consider MRI in a little more detail.

MRI uses a magnetic field and radio waves to image soft tissues. When placed in a magnetic field the nuclei of atoms line up with the direction of the magnetic field, in much the same way as a compass needle aligns itself to the Earth's magnetic field. Hydrogen atoms in water are monitored in MRI imaging because there is a high water content in the tissues under investigation and hydrogen has a strong tendency to line up with the magnetic field.

> ### Activity
>
> In **Activity 8.9** you use brain images to identify the different regions of the brain and their functions. Visit some of the websites associated with the activity to see lots of CT and MRI scans. **A8.09S**

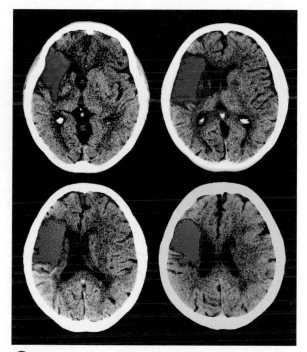

🔺 **Figure 8.33** Four CT images of the brain of a patient who has had a stroke. A massive brain lesion in the left hemisphere affects the language areas. Surrounding areas are still intact, allowing the patient to sing but not to speak.

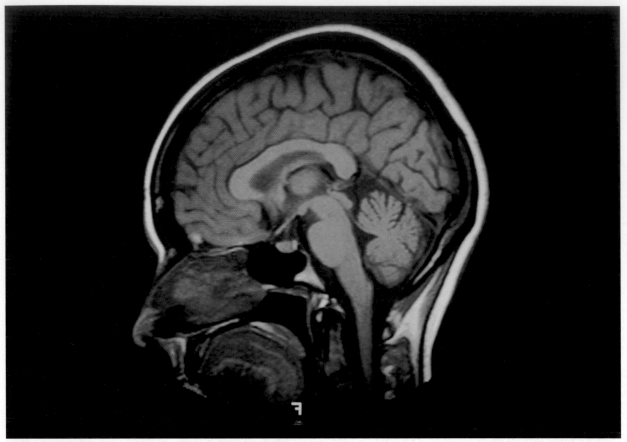

▲ **Figure 8.34** A typical MRI image showing good resolution of soft tissues.

In an MRI scanner a magnetic field runs down the centre of the tube in which the patient lies. Another magnetic field is superimposed on this, which comes from the magnetic component of high-frequency radio waves. The combined fields cause the direction (axis) and frequency of spin of the hydrogen nuclei to change, taking energy from the radio waves to do so. When the radio waves are turned off, the hydrogen nuclei return to their original alignment and release the energy they absorbed. This energy is detected and a signal is sent to a computer which analyses it to produce an image on the screen (Figure 8.34).

Different tissues respond differently to the magnetic field from the radio waves, and so produce contrasting signals and distinct regions in the image. MRI examines tissues in small sections, normally thin 'slices', which can be put together to give three-dimensional images.

Nowadays MRI is widely used in the diagnosis of tumours, strokes, brain injuries and infections of the brain and spine. MRI can be used to produce finely detailed images of brain structures, as shown in Figure 8.34, with better resolution than CT scans for the brain stem and spinal cord.

Functional magnetic resonance imaging

Functional magnetic resonance imaging (**fMRI**) is a particularly useful and exciting tool for the neuroscientist, as it can also provide information about the brain in action. Using this technique it is possible to study human activities such as memory, emotion, language and consciousness.

Extension

Use **Extension 8.4** to find out in more detail how brain-imaging techniques work and learn about some new techniques. **X8.04S** fMRI is used to look at the functions of the different areas of the brain by following the uptake of oxygen in active brain areas. This is possible as deoxyhaemoglobin absorbs the radio wave signal, whereas oxyhaemoglobin does not. Increased neural activity in a brain area results in an increased demand for oxygen, and hence an increase in blood flow. Although there is a slight increase in oxygen absorption from the blood, overall there is a large increase in oxyhaemoglobin levels in the enhanced blood flow so less signal is absorbed. The less radio signal that is absorbed, the higher the level of activity in a particular area, so different areas of the brain will 'light up' when they are active (Figure 8.35). You can see some fMRI images on the websites linked to Activity 8.9.

fMRI can produce up to four images per second, so the technique can be used to follow the sequence of events over quite short time periods. In a typical fMRI experiment, images are collected continually while the subject alternates between resting and carrying out some task, such as object recognition, listening or memorising number sequences.

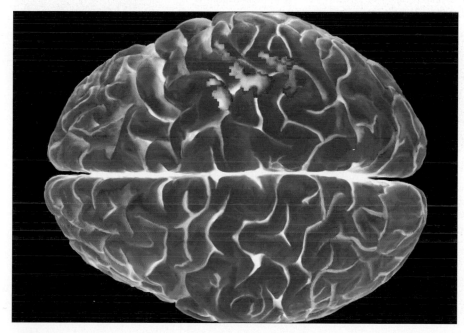

Figure 8.35 Active areas of the brain appear as coloured areas on a fMRI scan due to the high levels of oxyhaemoglobin present.

Q8.19 A neuroscientist conducts an fMRI experiment to investigate brain activity when subjects perform voluntary actions such as pressing a lever. Look back at the section on regions of the brain and decide which part of the brain would be expected to be active when the lever is being pressed.

Q8.20 A study of a group of London taxi drivers showed that the right hippocampus is involved in recalling a well developed mental map of London. A second study examined brain scans of 16 London taxi drivers and found that the only areas of their brains that were different from 50 control subjects were the left and right hippocampus. A particular region, the posterior hippocampus, was found to be significantly larger in the taxi drivers, whilst the front of the hippocampus was smaller than in the control subjects.
a What imaging method could have been used in: **i** the first and **ii** the second investigation?
b What do you think the scientists were able to infer from their results?

8.4 Visual development

How does the visual cortex develop? What about the connections from the eye to the visual cortex? Was Kenge's visual development in some way faulty, causing him to see the buffalo as insects?

The human nervous system begins to develop soon after conception. By the 21st day a neural tube has formed. The front part of the neural tube goes on to develop into the brain, while the rest of the neural tube develops into the spinal cord. The rate of brain growth during development is truly astonishing. At times, 250 000 neurones are created every minute!

A baby arrives in the world with about 100 000 million neurones. There is no huge increase in number of brain cells after birth, though there is a large postnatal increase in brain size. This is caused by several factors, principally elongation of axons, myelination and the development of synapses. By six months after birth the brain will have grown to half its adult size. By the age of two years, the brain is about 80% of the adult size.

Once neurones have stopped dividing, the immature neurones migrate to their final position and start to 'wire themselves'. Axons lengthen and synapse with the cell bodies of other neurones. Neurones must make the correct connections in order for a function such as vision to work properly.

Axon growth

Axons of the neurones from the retina grow to the thalamus. Here they form synapses with neurones in the thalamus in a very ordered arrangement. Axons from these thalamus neurones grow towards the visual cortex in the occipital lobe.

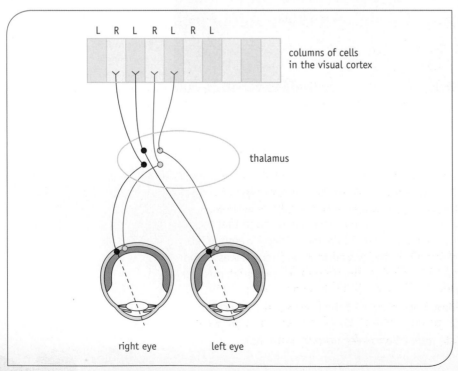

columns of cells in the visual cortex

thalamus

right eye   left eye

◀ **Figure 8.36** The relationship between the cells in each retina and the cells in the visual cortex.

Staining techniques and studies using electrical stimulation show that the visual cortex is made of columns of cells. Axons from the thalamus synapse within these columns of cells. Adjacent columns of cells receive stimulations from the same area of the retina in the right and left eye with the pattern repeated across the visual cortex (see Figure 8.36). In this way a map of the retina is created within the visual cortex.

It used to be thought that these columns were formed during a **critical period** for visual development after birth, the result of nurture rather than nature. This is now known not to be the case. Working with ferrets, Crowley and Katz showed by injecting labelled tracers that the columns in the visual cortex are formed before the critical period for development of vision (Figure 8.37). Columns are also seen in newborn monkeys, suggesting that their formation is genetically determined and not the result of environmental stimulation.

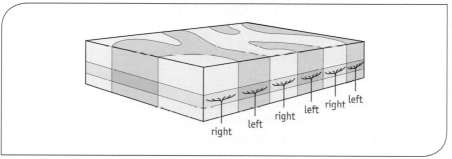

▲ **Figure 8.37** Radioactive label moves from one eye and is concentrated into distinct bands in the visual cortex showing the columns of cells that receive input from that eye. These banding patterns have been observed in animals that have received no visual stimulation.

However, periods of time during postnatal development have been identified when the nervous system must obtain specific experiences to develop properly. These are known as critical periods, **critical windows** or sensitive periods.

Evidence for a critical period in visual development

The way in which the environment affects the wiring of the nervous system has been extensively studied using the visual system. The evidence for a critical period in visual development comes from several sources, including medical observations of people with eye conditions affecting sight and results of experiments using animals.

Medical observations

One well known case is that of a young Italian boy who, as a baby, had a minor eye infection. As part of his treatment, his eyes were bandaged for just two weeks. When the bandage was removed he was left with permanently impaired vision.

Studies of people born with **cataracts** have also contributed to our understanding of critical periods in development. You may know that a cataract is the clouding of the lens of the eye (Figure 8.38); this affects the amount of light getting to the retina. If cataracts are not removed before the child is ten years old, they can result in permanent impairment of the person's ability to perceive shape or form, including difficulties in face recognition.

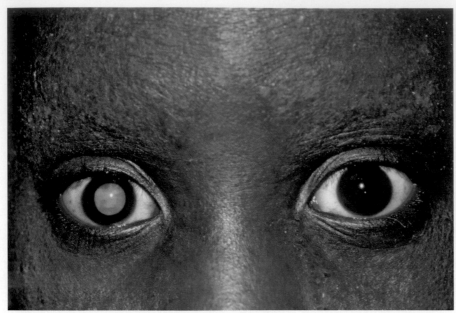

▲ **Figure 8.38** A cataract is the result of clouding of the lens.

By contrast, elderly people who develop cataracts later in life and then have them for several years report normal vision after their removal. This suggests that there is a specific time in development when it is crucial for light stimuli to enter the eye. Cataract removal in children is now carried out at a much earlier stage of development.

Research using animal models

Much of our knowledge and understanding of the visual system, and evidence for critical periods, comes from studies using animals. Animals are used extensively in the study of biological processes including brain development and function. Because most research is conducted on just a few types of animals, a wealth of information is available about them. They are known as **animal models**.

Most animal models are easy to obtain, easy to breed, have short life cycles, and have a small adult size. In Topic 3 we saw fruit flies, *Drosophila*, being used for studying the links between genes and development. Nematode worms, chickens, mice, frogs and zebrafish are also used in this area of research. Mice are used extensively in the study of cancer and disease. In the study of visual development kittens and monkeys have been used, because of their similarity to humans.

Research using animals raises ethical issues – see pages 231–3.

Studies of newborn animals

In one study, one group of newborn monkeys was raised in the dark for the first three to six months of their lives, and another was exposed to light but not to patterns. When the monkeys were returned to the normal visual world, researchers found that *both* groups had difficulty with object discrimination and pattern recognition.

Q8.21 What does this suggest is required for visual development in these monkeys?

In a series of studies, Hubel and Wiesel raised monkeys from birth to six months, depriving them of stimulus to light in one eye. This is known as **monocular deprivation**. After six months the eye was exposed to light. On exposure to light it was clear that the monkey was blind in the light-deprived eye. Retinal cells in the deprived eye did respond to light stimuli, but the cells of the visual cortex did not respond to any visual input from the formerly deprived eye. Deprivation for only a single week during a certain period after birth produced the same result, with the deprived eye failing to respond to light. Deprivation in adults had no effect. Interestingly, visual deprivation of both eyes during this critical window has much less effect than when just one eye is deprived.

Q8.22 Hubel and Wiesel tested kittens for the effects of monocular deprivation at different stages of development and for different lengths of time. They found:
- deprivation at under three weeks had no effect
- deprivation after three months had no effect
- deprivation at four weeks had a catastrophic effect – even if the eye was closed for merely a few hours.

Bearing in mind that kittens are born blind, can you explain the results above?

Q8.23 Young doves and chaffinches raised without exposure to adult song will sing the adult song perfectly the first time they try it – the behaviour is innate (inborn). Most other bird species need exposure to the adult song during a critical period in order to learn the proper song. Comment on the main influence on the development of the part of the brain associated with bird song: is it nature or nurture?

What happens during the critical period?

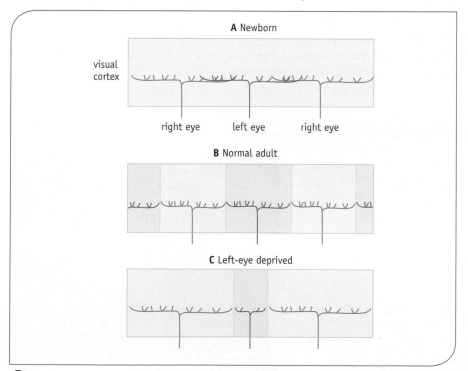

Activity

In **Activity 8.10** you can look at the evidence for a critical period for development of vision. **A8.10S**

▲ **Figure 8.39 A** Overlapping columns in the visual cortex are present at birth. **B** Refinement during the critical period produces the distinctive pattern of columns driven by the left and right eye. **C** Columns that receive input from a light-deprived eye become much narrower.

If the columns in the cortex are created before birth what is happening during the critical period which can result in impaired vision if one eye is deprived of light? There must be further visual development.

At birth in monkeys there is a great deal of overlap between the territories of different axons (Figure 8.39A). In adults, there is less overlap (Figure 8.39B) even though the mass of the brain is greater with more dendrites and synapses. Are these changes the result of visual experience or are they genetically determined?

After light deprivation in one eye, columns with axons from the light-deprived eye are narrower than those for the eye receiving light stimulation (Figure 8.39C). Dendrites and synapses from the light-stimulated eye take up more territory in the visual cortex. This suggests that visual stimulation is required for the refinement of the columns, and so for full development of the visual cortex.

Axons compete for target cells in the visual cortex. Every time a neurone fires onto a target cell, the synapses of another neurone sharing the target cell are weakened, and they release less neurotransmitter. If this happens repeatedly, the synapses that are not firing will be cut back (Figure 8.40). When one eye is deprived of light, the axons from that eye will not be stimulated. In the area of the visual cortex where neurones from both eyes overlap, the synapses on axons receiving light stimulation fire. The synapses from the light-deprived eye will be weakened and eventually lost (Figure 8.39). Throughout the nervous system many more neurones are produced than are required, and numerous synapses and axons are pruned back. In the retina up to 80% of the original neurones die.

Q8.24 Rewrite these sentences in the correct order, to fit together the pieces of this jigsaw of ideas.
1 There is a lack of visual stimulation in one eye.
2 Inactive synapses are eliminated.
3 Axons from the non-deprived eye pass impulses to cells in the visual cortex.
4 Synapses made by active axons are strengthened.
5 Axons from the visually deprived eye do not pass impulses to cells in the visual cortex.

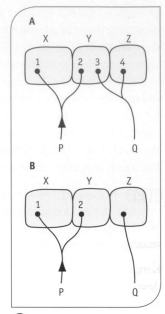

🔺 **Figure 8.40 A** Impulses passing along neurone P cause synapses 1 and 2 to fire. Synapse 3 releases less neurotransmitter and is eventually lost if there are many more impulses passing along P than Q. **B** Neurone Q no longer synapses with cell Y.

Did you know? Using stem cells to restore vision

Mike May (Figure 8.41), a successful businessman from San Francisco, lost his sight in a chemical explosion when he was three years old. Forty years later, in 2000, he underwent pioneering surgery on his cornea in a bid to restore some vision. Donor stem cells were transplanted into the eye. To be successful the cells needed to multiply and bind to the original cornea cells.

When the bandages came off, to his surprise a limited amount of vision had been restored. However, Mike's biggest problem is understanding what he's seeing. His main method of perception is still through touch. He has to learn how to interpret what he sees. He will never learn to see as clearly as most people.

🔺 **Figure 8.41** Mike May who is learning to see again.

8.5 Making sense of what we see

Individual neurones in the visual cortex respond in different ways to information from the retina, and to different characteristics of the object being viewed. Some neurones, called simple cells, respond to bars of light. Other cells, called complex cells, respond to edges, slits or bars of light that move; others to the angle of the edge; still others to contours, movement or orientation.

However, visual perception is not simply the creation of an image of what is being viewed, but involves prior knowledge and experience as the brain interprets the sensory information received from the retina to create our visual experience of the world.

Kenge could see the animals on the plain perfectly well, so his visual development was not a problem. Impulses were successfully sent from the retina to the cortex where cells were stimulated; he then had to make sense of what he saw. This seems to be where the problems arose; he failed to make sense of what he saw.

Pattern recognition

Recognising letters

Look at the list of letters in the margin. You should easily recognise the letters unless you have dyslexia. Some theories of perception suggest that a template for patterns is stored in our long-term memory. The template may not be an exact representation of the letters, rather a model that has all the key features. So when patterns are viewed their key features are matched with those in the long-term memory, allowing us to recognise and interpret what we see. There is still a lot about dyslexia that we don't know but it clearly has something to do with problems in recognising certain patterns, for example 'b' and 'd' may be confused.

CQOOCG

QQDCOQ

GGCOQD

CGQSSU

GCOQUC

ODQUCO

GHOQGD

DGOQDS

If you look quickly down the list of letters opposite you will probably be able rapidly to pick out the single letter H with its straight vertical and horizontal lines – the other letters all have curved features. You do not have to look at every single letter separately. Cells that respond to the horizontal and straight lines are used; once they have been stimulated the pattern is checked for the features of an H, using a stored template.

When reading we do not have to recognise every single letter. If we did it would take much longer to read this sentence, in the way a small child must sound out each letter in turn. Instead we must be recognising complete words.

Recognition of objects probably works in a similar way. Try Question 8.25 and see if you can recognise the objects. This also shows how the brain makes assumptions about what it expects to see, based on the context and past experience, to make a best estimate at recognition.

Q8.25 Look at the two pictures in Figure 8.42, and decide what each is.

Activities

In **Activity 8.11** and **Activity 8.12** you can investigate pattern recognition. **A8.11S** and **A8.12S**

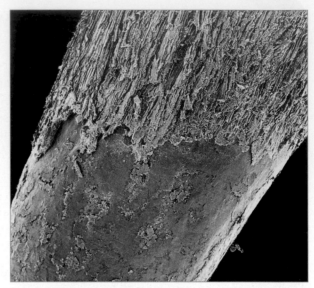

🔺 **Figure 8.42** Can you recognise these objects? The answers to Question 8.25 will let you check if you were correct.

Face recognition

Recent work has shown that we have specialised areas of cells, called recognition units, for identifying complex objects. One of these areas is specifically concerned with face recognition – it is called the face-recognition unit. (See the story of Lincoln Holmes on page 210.)

When we first see a new face, certain complex cells in the cortex are stimulated by the lines and contours of the face. These form synapses with other areas of the brain including the language and association areas in the brain. When we see the same face again, we may recognise it, as the necessary neural pathways are already there. In a well known face, only a few of the visual cues are needed to trigger recognition, as the system is already set up to respond. Look at the fragments of faces in Figure 8.43 and try Question 8.26 to check how many you recognise.

Q8.26 Name the owners of the faces in Figure 8.43. Explain why you might recognise some but not all of the individuals.

🔺 **Figure 8.43** Who are they? Check the answers to Question 8.26 to find out if you are correct.

During development, and indeed throughout our lives, we build up information in the brain that we call upon to interpret sensory information. Information in these libraries is reinforced (and easier to recall) the more times the information is put in. This is why reading your notes and revising thoroughly helps in your exam! You will learn about the molecular basis of this process later in this topic.

Depth perception

Kenge laughed when he was told that the animals on the horizon were buffalo. As they drove closer he recognised them as buffalo, so he must have known what a buffalo was. At some point he must have seen a buffalo up close and learned what it looked like, even though he could not perceive it as one from a distance. Was there a problem with his depth perception? Was he failing to realise that the object was far away and therefore would appear smaller?

When we look at any object we can make a judgement about how far away the object is. The brain does this in different ways for close and distant objects.

Close objects

For objects less than 30 m away from us, we depend on the presence of cells in the visual cortex that obtain information from both eyes at once. The visual field is seen from two different angles, and cells in the visual cortex let us compare the view from one eye with that from the other. This is called **stereoscopic vision** and it allows the relative positions of objects to be perceived.

Q8.27 Why might a child who had had an eye patch during visual development never develop stereoscopic vision?

Distant objects

For objects that are more than 30 m distant, the images on our two retinas are very similar, so visual cues and past experiences are used when interpreting images. Look at the beach in Figure 8.44. What visual cues might help in depth perception? Lines converge in the distance giving perspective, the impression of distance. The stones further away are smaller and, because from experience we know that the stones along a beach are usually roughly the same size, the small ones in the picture are perceived as being further away. We know the size of buildings, so the ones in the background are not smaller, just further away.

Overlaps of objects and changes of colour also help in judging depth. Even if an object becomes smaller on the retina, for example when a car drives away into the distance, we perceive it as moving further away not getting smaller.

Activity

Check if you have stereoscopic vision using **Activity 8.13**. A8.13S

◀ **Figure 8.44** Visual clues in the photograph allow depth perception.

Usually we are unaware of the cues being used. It is only when they are not present, or cause the brain to misinterpret the image, that we become aware of them, such as when viewing optical illusions – see Figure 8.45. The person standing further away may look taller, but the lines that imply depth fool the brain. If you measure the people with the ruler you will see that they are actually the same height.

Is depth perception innate or learned?

It is possible that Kenge had not developed depth perception over long distances because he had had no experience of seeing long distances on an open plain, having always lived in the enclosed forest. Studies into how people with different cultural experiences and how newborn children respond to optical illusions help answer this question.

▲ **Figure 8.45** Are these two people different heights? Check with a ruler to see if you are correct.

Cross-cultural studies

What is culture? It has been defined in a number of different ways. In this course we shall view culture as a system of beliefs that are shared among a group of people. It thus shapes experience and behaviour. People from different cultures may not share the same beliefs, and they may show different behaviours.

The Müller–Lyer illusion

The Müller–Lyer illusion may well be familiar to you. Look at Figure 8.46 and decide which of the two vertical lines, X or Y, is longer. The two lines are actually the same length, but most people mistakenly think they differ in length.

According to the **carpentered world hypothesis**, those of us who live in a world dominated by straight lines and right angles tend to interpret images with acute and obtuse angles as right angles. Thus a non-rectangular image, such as a trapezium (a four-sided figure with only two parallel sides), is perceived as rectangular (i.e. having four right angles of 90°). This interpretation of acute and obtuse angles (so long as they don't differ too greatly from 90°) becomes automatic and unconscious from a very early age.

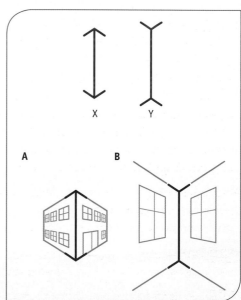

▲ **Figure 8.46** The Müller–Lyer illusion.

In the Müller–Lyer illusion the angles between the line and the arrow heads are interpreted as right angles providing depth cues to the image. It has been argued that line X is perceived as the outside corner of a building (Figure 8.46A), but line Y as the inside corner of a room (Figure 8.46B). Because buildings are normally viewed at a distance and the height of the nearest corner appears to be greater than the rest of the building, the brain shortens the length of line X. The height of the far corner of a room, on the other hand, appears to be less than the rest of the room, so the brain compensates by increasing the perceived length of line Y.

Studies have shown that people who live in a 'circular culture' with few straight lines or right angle corners, such as the Zulu people of Africa who have circular houses and no roads, rarely misjudge the length of the lines in the Müller–Lyer illusion. They have little experience of interpreting acute and obtuse angles on the retina as representations of right angles.

Some researchers think that lack of susceptibility to the Müller–Lyer illusion is not the result of different experience, but is due to genetic differences in pigmentation between individuals. They suggest that individuals who find it harder to detect contours are less susceptible to the Müller–Lyer illusion. They link poor contour detection to higher retinal pigmentation. In light-coloured people, with low retinal pigmentation, contour detection is good so such people are more easily caught out by the illusion.

Q8.28 Why might researchers be cautious about the pigment and contour-detection evidence supporting the idea that susceptibility to the Müller–Lyer illusion is genetic rather than environmental?

Activity

Investigate the Müller–Lyer illusion using the interactive **Activity 8.14. A8.14S**

Depth cues in other pictures

In another study, individuals from a range of different cultures were shown pictures with depth cues such as object size, overlap of objects and linear perspective, similar to those in Figure 8.47.

It was found that all young children had difficulty perceiving the pictures as three-dimensional. They would have said that the man in Figure 8.47 was pointing at the elephant not the antelope – they failed to interpret the depth cues. By the age of 11 years, almost all the European children interpreted the pictures in three dimensions, but some Bantu and Ghanaian children still did not, as did non-literate adults, both Bantu and European. They had had less experience in the interpretation of depth cues in pictures.

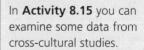 **Figure 8.47** Pictorial depth cues tell us that the man must be pointing at the antelope, not the elephant.

What does seem clear is that the depth cues in pictures which most of us take for granted are not **innate**; they have to be learned. It suggests that visual perception is, in part at least, *learned*.

Q8.29 What depth cues are used to decide that the man is pointing at the antelope in Figure 8.47 and not the elephant?

Q8.30 Suggest reasons why caution should be used when interpreting some of these cross-cultural studies.

Activity

In **Activity 8.15** you can examine some data from cross-cultural studies.

Studies with newborn babies

The fact that babies are born with a range of characteristic behaviours suggests that these are determined by genes; for example, crying, walking movements and grasping are all present from birth. Within 24 hours of birth newborn babies can distinguish human faces and voices from other sights and sounds, and prefer them. These types of inborn capacities exhibited by a newborn baby are used as evidence for the role of genes in determining the hard-wiring of the brain before birth.

Although born short-sighted, babies can see people and items clearly at a distance of about 30 cm. It is suggested that their preference for stripes and other patterns shows they are imposing order on their perceptions in early infancy. Long before they can crawl, they can tell the difference between a happy face and a sad one. They imitate people's expressions, and by the time they are old enough to pick up a phone they can mimic what they have seen others doing with it.

The visual cliff

In a classic experiment babies are encouraged to crawl across a table made of glass or Perspex below which is a visual cliff. Patterns placed below the glass create the appearance of a steep drop, as shown in Figure 8.48. If the perception of depth is innate, then babies should be aware of the drop even if they have not previously experienced this stimulus themselves.

Q8.31 If the baby does have a perception of depth, how will she react when invited to crawl over the 'edge' of the cliff?

Young babies were very reluctant to crawl over the 'cliff' even when their mothers encouraged them to do so.

Q8.32 **a** This reaction is assumed to indicate that depth perception is innate. Explain why.
b Argue the case for this not illustrating that depth perception is innate.

This experiment is only possible with babies who have learned to crawl. It is likely that a six-month-old human may already have learned depth perception. Therefore the experiment was repeated with animals that can walk as soon as they are born such as chicks, kids (young goats) and lambs. They too refused to cross the cliff.

Q8.33 Explain if this supports or does not support the idea that this type of perception is innate.

Q8.34 Having read Section 8.5, can you suggest why Kenge, a forest dweller, was fooled into thinking that the buffalo on the horizon were insects?

▲ **Figure 8.48** The visual cliff.

8.6 Learning and memory

Kenge knew what buffalo looked like even though he did not perceive them correctly from a distance. He had learned what they looked like and could recall this information from his memory. What is learning and how do we store information in our memory?

Learning is any relatively permanent change in behaviour or knowledge that comes from experience, and it occurs throughout your life. For learning to be effective, you must be able to remember what you have learned, and studies of learning and memory have always gone hand in hand. Throughout our lives, the memory stores vast amounts of information, from sights and sounds to emotions and skills such as riding a bike or texting. But what is learning and how are memories stored? Some classic studies of learning illustrate the different types of learning.

Types of learning

Classical conditioning

When the bell next goes for lunch, think about your mouth. Is it watering? Although you probably had not realised it, you may well have learned to associate the sound of the bell with the arrival of food. This means that the stimulus of the bell promotes a response to produce saliva. The psychologist Ivan Pavlov studied these type of associations with dogs (see Figure 8.49).

When Pavlov's dogs successfully associated the sound stimulus with being fed, the dogs' behaviour changed relatively permanently. This meant that learning had taken place. The sound of the bell is known as a **neutral stimulus** because it would not have promoted salivation before the experiment. The following process has occurred:

1 The dog responds to an **unconditioned stimulus** (receiving food).

2 The neutral stimulus (sound) becomes associated with the unconditioned stimulus (receiving food).

3 The dog responds to the neutral stimulus whether or not food is present.

This is known as a **conditioned response**. This type of learning is known as **classical conditioning**.

Q8.35 Name the neutral stimulus, unconditioned stimulus and conditioned response in these examples:
a the Pavlov dog experiment
b an experiment in which air was puffed into a person's eye at the same time that a sound was heard; eventually the sound alone caused blinking.

Pavlov's dogs showed how conditioning occurs when the neutral stimulus immediately precedes the unconditioned stimulus. If the presentation of food is just as likely with or without the sound preceding it, then no association will develop, or any existing association will be suppressed.

▲ **Figure 8.49** Pavlov made a sound just before each occasion when the dog received food. The dog learned to associate the sound with food and would produce saliva on hearing the sound even if no food followed.

Activity

In **Activity 8.16** you can investigate conditioning, including conducting Pavlov's dog experiment yourself, using the animation on the Nobel Prize website given in the weblinks that accompany this activity. **A8.16S**

Once Pavlov's dogs were conditioned to produce saliva on hearing the sound of a bell, they had learned to associate two stimuli – food and the sound. When one neurone is stimulated, for example by the sight of food, a neurotransmitter, frequently glutamate, is released. This binds to receptors on the postsynaptic membrane, allowing sodium ions to flood in and generate an action potential in the postsynaptic cell. If a second stimulus occurs at the same time from a second synapse, for example due to the sound of a bell, the second synapse is strengthened. The postsynaptic neurone has already been stimulated, so it responds more strongly because a second set of channels is able to open. This is due to the depolarisation by the previous impulse, allowing more sodium ions to enter the cell. In the case of Pavlov's dogs, this strengthened the link between the sound and salivation. It is thought that this is what happens when synapses are strengthened in the development of the visual cortex.

Operant conditioning

Classical conditioning is not the only form of conditioning. Jayshree was walking along the top of a narrow wall one day, when she lost her footing and fell off. She hurt herself slightly, and her mother shouted 'That'll teach you'. Jayshree can learn from the consequences of her behaviour. This type of learning, illustrated in the experiment shown in Figure 8.50, is an example of **operant conditioning**, where the consequences of an animal's behaviour affect whether that behaviour will occur again. The reward is a **reinforcement**. The reinforcement makes it more likely that the animal will repeat the behaviour. The cat in Figure 8.50 learns to press the lever to escape from the box.

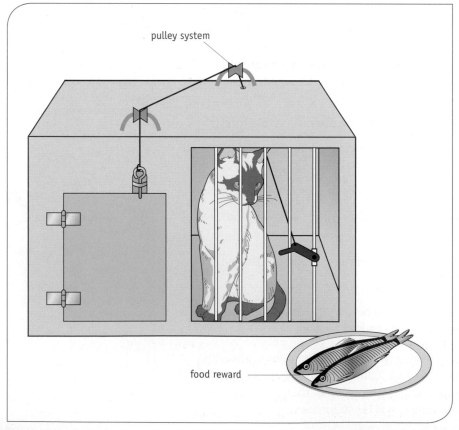

pulley system

food reward

◀ **Figure 8.50** In a classic set of experiments, Edward Thorndike (1898) placed a cat in a box with a plate of fish outside. To get out of the box, the cat had to press a latch, which opened the door. By a process of trial and error, the cat learned how to escape from a box. In the early stages, the cat simply made lots of random movements, eventually operating the door latch by accident. Gradually, the cat began to associate the particular movement of operating the latch with food. Having made this association, it could open the door very quickly.

Punishments work in a similar way. If the animal received an electric shock when it pressed the lever, it would soon learn to press less. It would learn that pressing the lever is associated with pain. In the same way Jayshree will learn that falling is painful and she will learn to be more careful.

Did you know? Training animals using reinforcement

By using operant conditioning, you can teach an animal remarkably complex behaviours. Imagine you want a rat to stretch up and push a button on the roof of a box with its nose. It is quite unlikely the rat would ever do this by chance. So how do you use reinforcement to achieve your aim? The best method is to break the desired behaviour into stages. For example:

1 Reward the (male) rat every time he stretches up, until he is stretching up quite often.

2 Stop rewarding for stretching, and reward him every time a part of his body touches the button.

He should soon be stretching up and touching the button frequently.

3 Stop rewarding for general touching, and reward him only when his nose touches the button. He should soon be touching the button specifically with his nose.

4 Stop rewarding for just touching, and only reward him when his nose pushes the button. He should soon frequently be pushing the button with his nose: the desired behaviour.

Insightful learning

Wolfgang Köhler conducted experiments on chimpanzees. He placed a banana outside a cage, just out of reach. The chimp saw the banana and sat down. A short time later, it got up, picked up two sticks, joined them together and reached for the banana. Köhler called this type of learning **insightful learning**.

Learning by observing

Learning by observing involves watching others and copying them (or deciding not to copy them). If your teacher/lecturer shows you how to do a practical by demonstrating it, you learn by watching them. Learning by observing does not appear to involve classical or operant conditioning. If learning by observing did not take place, many things would take a lot longer to learn.

Types of memory

When something is learned it is stored in the memory. There are three main ways information can be stored: semantic (meanings), visual (pictures) and auditory (sounds). For example, the word 'bike' could be stored as a picture or visual image of the word, as a sound by saying the word aloud or as a meaning related to the word such as when you ride the bike. It can be stored in the short-term memory and then the long-term memory.

Short-term memory

The short-term memory has a limited capacity. Many researchers believe it can hold only between five and nine different pieces of unconnected information at once. Estimates on how long information remains in short-term memory range from 4 seconds to a couple of minutes. Indeed, you can probably not remember the exact wording of the top line of this paragraph, even though you read it only about 10 seconds ago!

To keep information in short-term memory (and therefore to give it more chance of passing into long-term memory) you can rehearse that information. For example, when you try to remember a phone number and say it over and over again to yourself, you are using rehearsal to keep the number in your short-term memory. Speaking it out loud or writing it down helps because it is thought that for most people auditory and visual information are more readily stored in short-term memory than semantic information.

Short-term memory also acts as a location to which information from long-term memory can be brought and used. For this reason it is often called the working memory. When doing mental arithmetic, for example, the instructions about what to do with the numbers arrive from the long-term memory, where they were stored when you first learned to add and subtract.

Long-term memory

Long-term memory differs from short-term memory in that there are two types of memories in long-term memory:

- memories about facts or events, involving 'knowing that ...'
- memories that we have, but are not consciously fully aware of, involving 'knowing how to ...'. These memories allow us to learn motor skills such as writing and walking; once learned they are rarely forgotten.

Recall from long-term memory is most effective when we store that information according to its meaning; linking information to previously stored information provides a good cue for recalling it later.

Where memories are stored

Different types of memory are controlled by different parts of the brain. This is clearly demonstrated by looking at cases where people have lost the use of particular parts of the brain, as the case study below illustrates.

In a patient, HM, who suffered from severe epilepsy, doctors in 1953 removed those areas of the brain that appeared to be causing the problem, and immediately caused amnesia. HM's long-term memories from before the operation were unaffected, but he could no longer form new long-term memories, and found it difficult to remember what he had done just a few minutes before. By contrast, his memory for how to do everyday things was still intact.

Q8.36 Look back at the section on regions of the brain and find out which parts of the brain are involved in memory.

How memories are stored

In the brain, every neurone connects with many other neurones to make up a complex network. Memories can be created in two ways, by altering:

- the pattern of connections
- the strength of synapses.

Activity

In **Activity 8.17** you can learn a new skill, using rehearsal to fix it in your memory. **A8.17S**

Extension

Extension 8.5 investigates which type of information is most successfully stored in short-term memory and the capacity of your short-term memory. **X8.05S**

Sea slugs and habituation – changing synapse strength

◀ **Figure 8.51** The giant sea slug (*Aplysia californica*). This species may grow to be 30 cm in length and weigh 1 kg.

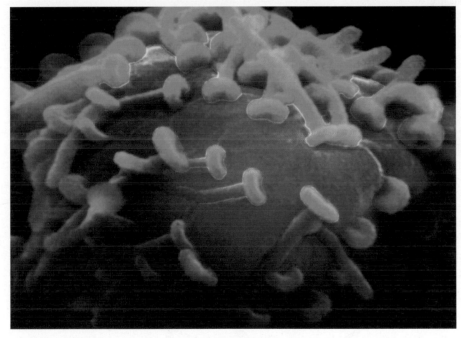

◀ **Figure 8.52** Scanning electron micrograph of synapses in the giant sea slug. Magnification ×2350.

Eric Kandel, amongst others, studied the molecular biology of learning in the giant sea slug (*Aplysia*, Figure 8.51) to help understand learning in humans. He shared a Nobel Prize in 2000 for his work on *Aplysia*.

There are no fundamental differences between the nerve cells and synapses of humans and those of animals such as the sea slug. However, with only 20 000 neurones, the neurobiology of a sea slug is much simpler than that of a human. Sea slugs have large accessible neurones (Figure 8.52) so those involved in particular behaviours can be identified. Sea slug behaviour can be modified by learning, and the effects on neurones and synapses studied.

Q8.37 **a** Sea slug neurones cannot pass impulses by saltatory conduction. Suggest a reason for this.

b Suggest how the sea slug neurones can transmit impulses rapidly.

The giant sea slug breathes through a gill located in a cavity on the upper side of its body. Water is expelled through a siphon tube at one end of the cavity. If the siphon is touched, the gill is withdrawn into the cavity (Figure 8.53). This is a protective reflex action similar to removing a hand from a hot plate.

Because they live in the sea, *Aplysia* are frequently buffeted by the waves. However, they learn not to withdraw their gill every time a wave hits them. They become habituated to the waves. **Habituation** is a type of learning. When you put your socks on you feel them at first, but after a few minutes you no longer notice them. You have become habituated to the feeling of the socks on your feet, even though they are still providing a touch stimulus.

Habituation allows animals to ignore unimportant stimuli so that limited sensory, attention and memory resources can be concentrated on more threatening or rewarding stimuli.

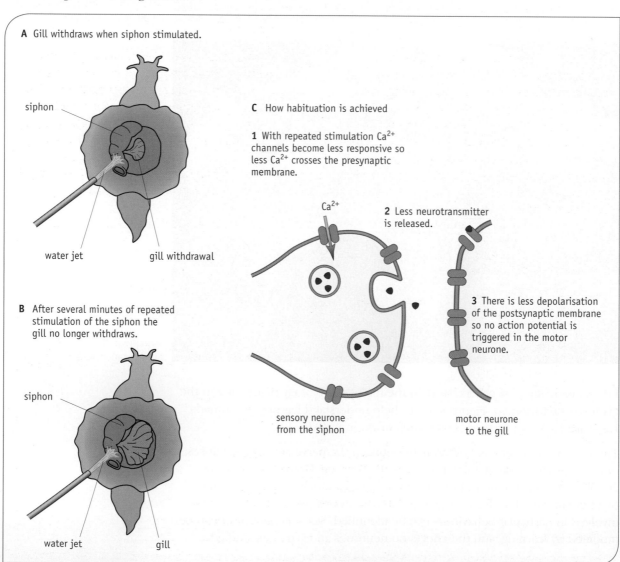

A Gill withdraws when siphon stimulated.

siphon

water jet   gill withdrawal

B After several minutes of repeated stimulation of the siphon the gill no longer withdraws.

siphon

water jet   gill

C How habituation is achieved

1 With repeated stimulation Ca^{2+} channels become less responsive so less Ca^{2+} crosses the presynaptic membrane.

Ca^{2+}

2 Less neurotransmitter is released.

3 There is less depolarisation of the postsynaptic membrane so no action potential is triggered in the motor neurone.

sensory neurone from the siphon

motor neurone to the gill

🔺 **Figure 8.53** Stimulating the siphon causes the gill of the sea slug to be withdrawn.

Q8.38 Sketch a diagram to show the reflex arc that is involved in withdrawal of the sea slug gill.

What happens during habituation?

Kandel stimulated the sea slug's siphon repeatedly with a jet of water. The response gradually faded away, until the gill was not withdrawn any more. The neurones involved in the reflex were identified, and Kandel found that the amount of neurotransmitter crossing the synapse between the sensory and motor neurones decreased with habituation. With repeated stimulations, fewer calcium ions move into the presynaptic neurone when the presynaptic membrane is depolarised by an action potential; fewer neurotransmitter molecules are then released (Figure 8.53).

More connection – longer memory

Long-term memory storage can also involve an increase in the number of synaptic connections. Repeated use of a synapse leads to creation of additional synapses between the neurones.

> **Activity**
>
> In **Activity 8.18** you can investigate what happens at synapses during habituation. **A8.18S**

Did you know? Sensitisation

Sensitisation is the opposite of habituation. It happens when an animal develops an enhanced response to a stimulus. Humans can learn by sensitisation. Imagine you are at home alone late at night, and you hear a loud crash from outside. For a short period after the crash, any small noises (which you previously had not noticed because you were habituated to them) seem very loud and provoke a similar enhanced response. You have become sensitised to these noises.

If a predator attacks *Aplysia*, the sea slug becomes sensitised to other changes in its environment and responds strongly to them. In experiments studying sensitisation Eric Kandel gave an electric shock to the sea slug tail before stimulating the siphon again with a jet of water. This provoked an enhanced gill withdrawal response that lasted from several minutes to over an hour. Use Figure 8.54 to help you to understand how sensitisation occurs.

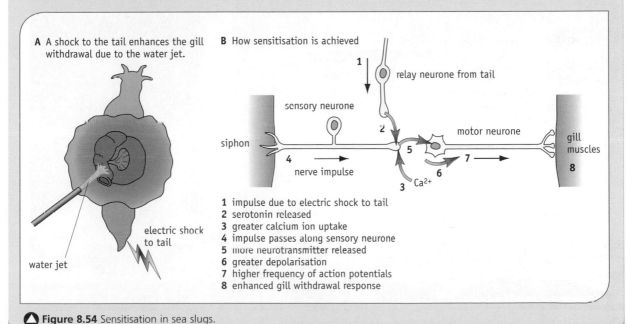

1 impulse due to electric shock to tail
2 serotonin released
3 greater calcium ion uptake
4 impulse passes along sensory neurone
5 more neurotransmitter released
6 greater depolarisation
7 higher frequency of action potentials
8 enhanced gill withdrawal response

Figure 8.54 Sensitisation in sea slugs.

The ethics of using animals in medical research

The work of Kandel and others on sea slugs has helped in our understanding of learning. However, people vary greatly in their views about whether or not it is acceptable to use animals in medical research (Figure 8.55) – or for other purposes such as farming or as pets.

First of all there are those who believe in **animal rights**. People don't any longer consider it morally acceptable to have human slaves so, the argument goes, why is it acceptable to keep animals captive in laboratory cages or on farms? If you believe that humans have certain rights, it is quite difficult to explain why animals have none.

Of course, this doesn't mean that animals such as chimpanzees or dogs would have a right to vote or join a trade union – such rights are as meaningless for them as they are for a three-year-old child. But giving laboratory, domestic and farm animals rights would mean that they should be entitled to such things as food, water, exercise and veterinary treatment.

The importance of consent

From the point of view of medical research, accepting that animals have rights would mean that we could use only animals that consented to participate in medical experiments, just as we only use humans in medical experiments if they give their informed consent. In practice, except perhaps for such things as feeding trials, this would bring an end to the use of animals in medical research.

⬥ **Figure 8.55** The use of animals in medical research is strictly regulated in the UK and many other countries, but remains controversial.

Animal welfare rather than animal rights

A much more widespread position than believing that animals have rights is the belief that humans should treat animals well so far as is possible. Here the emphasis is on **animal welfare**. This is pretty much the position in European law. No country in the European Union is allowed to use vertebrates in medical experiments if there are non-animal alternatives. If there are not, animals can be used provided strict guidelines are followed.

Animal suffering and experience of pleasure

Both the animal rights approach and the animal welfare approach assume that animals can suffer and experience pleasures. This seems pretty obvious to anyone who has had a pet cat or dog. But what of fish? Can they suffer or experience pleasures? There is genuine scientific and philosophical debate – though the balance is turning in favour of the fish. And what of spiders and insects? Most experts reckon they cannot suffer. If you spend your weekends pulling the wings off flies it probably means that you are a most inconsiderate and unpleasant person (or have suffered an abusive childhood); but it probably doesn't increase the amount of animal suffering in the world.

A utilitarian approach to the use of animals

One ethical framework we have introduced in this course is **utilitarianism** – the belief that the right course of action is one that maximises the amount of overall happiness or pleasure in the world (see Topic 2). A utilitarian framework allows certain animals to be used in medical experiments *provided* the overall expected benefits are greater than the overall expected harms. Suppose, for example, to oversimplify greatly, that it takes the lives of 250 000 mice used in medical experiments to find a cure for breast cancer and that 50 000 of these mice are in pain for half their lives. There could still be a utilitarian argument for using the mice.

Q8.39 What would be the utilitarian argument for using 250 000 mice to find a cure for breast cancer?

Activity

In **Activity 8.19** you investigate what people think of using animals for medical research.
A8.19S

8.7 Problems with synapses

The functioning of the nervous system is completely dependent on synapses passing impulses. In everything we do, with every sensation, action and thought, neural pathways are stimulated. The control and coordination of all this activity in the brain relies on synapses. The synapses in turn depend on the neurotransmitters. Unfortunately things can go wrong, with adverse consequences for health. Imbalances in naturally occurring brain chemicals can cause problems, as can drugs when they cross the blood-brain barrier. The endothelial cells forming the capillaries in the brain are more tightly joined together than those in capillaries elsewhere in the body. They form a barrier to control the movement of substances into the brain and protect the brain from the effects of changes in blood ionic composition and from toxic molecules.

There are about 50 different neurotransmitters in the human central nervous system, and they need to be present in controlled amounts. The function of two examples, dopamine and serotonin, are discussed here, together with the consequences of abnormal levels. The effects of ecstasy illustrate how drugs can affect synapses.

Activity

In **Activity 8.20** you can refresh your memory of how synapses work, and review how they are affected by chemical imbalances. **A8.20S**

Parkinson's disease

Dopamine and Parkinson's disease

Dopamine is a neurotransmitter secreted by neurones including many located in part of the midbrain. The axons of the neurones in this area extend throughout the frontal cortex, the brain stem and the spinal cord. They are important in the control of muscular movements. Dopamine has also been linked with emotional responses.

In people with Parkinson's disease, dopamine-secreting neurones in the basal ganglia die, which means there is reduced dopamine in the brain. The main symptoms of the disease are:

- stiffness of muscles
- tremor of the muscles
- slowness of movement
- poor balance
- walking problems.

Other problems that may arise include depression and difficulties with speech and breathing. Between 1% and 2% of people in the UK over 50 are affected by Parkinson's disease, although the onset can be well before that age.

Treatment for Parkinson's disease

Recent developments in drug treatments for Parkinson's have made it much easier for some people to live with the effects of the disease. Some of the treatments for Parkinson's disease are outlined below.

- Slowing the loss of dopamine from the brain, with the use of drugs such as selegiline. This drug inhibits an enzyme called monoamine oxidase, which is responsible for breaking down dopamine in the brain.

- Treating the symptoms with drugs. Dopamine itself cannot be given to treat Parkinson's because it cannot cross into the brain from the bloodstream, but L-dopa, a precursor in the manufacture of dopamine, can be given. Once in the brain L-dopa is converted into dopamine, increasing the concentration of dopamine and controlling the symptoms of the disease.
- Use of dopamine **agonists**. Dopamine agonists are drugs that activate the dopamine receptor directly. These drugs mimic the role of dopamine in the brain, binding to dopamine receptors at synapses and triggering action potentials. They can be particularly useful in the treatment of Parkinson's disease, since they avoid higher than normal levels of dopamine in the brain. Abnormally high dopamine levels can have unpleasant side effects (see Did you know? below).
- New surgical approaches are being trialled, some of which are generating encouraging results.

> **Weblink**
>
> Visit the Birmingham Department of Clinical Neuroscience website to see some Parkinson's video clips showing the effect of treatment with L-dopa.

Did you know? Schizophrenia, hyperactivity and cocaine

Excess dopamine in the brain is believed to be a major cause of schizophrenia. This excess dopamine can be treated with drugs that block the binding of dopamine to its postsynaptic receptor sites. These drugs are usually similar to dopamine in structure, but unable to stimulate the receptors. This reduces the effect of the dopamine in triggering postsynaptic action potentials. A side effect in patients taking these drugs is to induce the symptoms of Parkinson's disease.

Ritalin is a drug used to treat childhood hyperactivity associated with low levels of dopamine. The effect of the drug is to prevent dopamine from being taken back up by the presynaptic membrane; more dopamine is left in the synaptic cleft to stimulate postsynaptic receptor molecules.

Cocaine also prevents dopamine reuptake, by binding to proteins which normally transport dopamine. Not only does cocaine bind to the transport proteins in preference to dopamine, it remains bound for much longer than dopamine. As a result, more dopamine remains to stimulate neurones; this causes prolonged feelings of pleasure and excitement. Amphetamines also increase dopamine levels. Again, the result is over-stimulation of these pleasure-pathway nerves in the brain. Long-term amphetamine use can induce schizophrenia-like illness.

Depression

Serotonin and depression

The neurotransmitter **serotonin** plays an important part in determining a person's mood. In the brain the groups of neurones that secrete serotonin are situated in the brain stem. Their axons extend into the cortex, the cerebellum and the spinal cord. Neurones that release serotonin have targets in a huge area of the brain (Figure 8.56).

A lack of serotonin has been linked to depression. Clinical depression is a relatively common condition, involving far more than feeling a bit glum. It can last for months or years and can have a profound effect on work and relationships. It is under-diagnosed, particularly in children. Depression is associated with feelings of sadness, anxiety and hopelessness. Loss of interest in pleasurable activities and low energy levels are common, as are insomnia, restlessness and thoughts of death.

The causes of depression are not completely understood. There may be a genetic element, since it runs in families. However, for most people it probably has more to do with environment and upbringing than genes. It is probable that a number of factors are involved. Fewer nerve impulses than normal are transmitted around the brain, which may be related to low levels of neurotransmitters being produced. A number of neurotransmitters may have a role to play in depression, including dopamine and noradrenaline, but reduced serotonin levels seem to be most commonly involved.

Pathways involving serotonin have a number of irregularities in people with depression. The molecules needed for serotonin synthesis are often present in low concentrations, but serotonin-binding sites are more numerous than normal, possibly to compensate for the low level of the molecule.

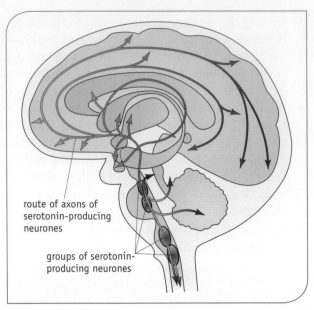

route of axons of serotonin-producing neurones

groups of serotonin-producing neurones

🔺 **Figure 8.56** The neurones that secrete serotonin are located in the brain stem.

Treatment for depression

One of the most effective types of drug involved in treating the symptoms of depression inhibits the reuptake of serotonin from synaptic clefts. A drug of this type is a selective serotonin reuptake inhibitor (SSRI), meaning that it blocks only the uptake of serotonin. One of the more common drugs of this type is Prozac, which maintains a higher level of serotonin, and so increases the rate of nerve impulses in serotonin pathways. This has the effect of reducing some of the symptoms of depression.

How drugs affect synaptic transmission

Synapses have a number of features that can be disrupted by interference from certain chemicals (Figure 8.57). For example, a chemical with a similar molecular structure to a particular neurotransmitter is likely to bind to the same receptor sites, and perhaps stimulate the postsynaptic neurone. Other chemicals may prevent the release of neurotransmitter, block or open ion channels, or inhibit a breakdown enzyme such as acetylcholinesterase. In all of these cases the normal functioning of the synapse will be disrupted, with consequences that depend on the nerve pathway involved.

The effect of ecstasy

The drug known as **ecstasy** affects thinking, mood and memory. It can also cause anxiety and altered perceptions (similar to but not quite the same as hallucinations). The most desirable effect of ecstasy is its ability to provide feelings of emotional warmth and empathy.

Ecstasy is a derivative of amphetamine. Its chemical name is 3,4-methylenedioxy-N-methylamphetamine (MDMA). The short-term effects of ecstasy include changes in brain chemistry and behaviour. The long-term effects include changes in brain structure and behaviour.

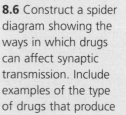

Figure 8.57 Five different stages in synaptic transmission that can be affected by drugs. Ecstasy binds to serotonin transport molecules preventing uptake back across the presynaptic membrane.

The diagram shows:
- **1** neurotransmitter synthesis and storage
- presynaptic membrane
- **2** neurotransmitter release
- **5** neurotransmitter breakdown
- **4** neurotransmitter reuptake
- postsynaptic membrane
- **3** neurotransmitter–receptor binding

How ecstasy affects synapses

Ecstasy increases the concentration of serotonin in the synaptic cleft. It does this by binding to molecules in the presynaptic membrane which are responsible for transporting the serotonin back into the cytoplasm. This prevents its removal from the synaptic cleft. The drug may also cause the transporting molecules to work in reverse, further increasing the amount of serotonin outside the cell. These higher levels of serotonin bring about the mood changes seen in users of the drug. It is possible that the ecstasy has a similar effect on molecules that transport dopamine as well.

Side effects of using ecstasy

Users report feelings of euphoria, well-being and enhanced senses. There are side effects and unpredictable consequences of using the drug. Some people experience clouded thinking, agitation and disturbed behaviour. Also common are sweating, dry mouth (thirst), increased heart rate, fatigue, muscle spasms (especially jaw-clenching) and hyperthermia, as ecstasy can disrupt the ability of the brain to regulate body temperature. Repeated doses or a single high dose of ecstasy can cause hyperthermia, high blood pressure, irregular heartbeat, muscle breakdown and kidney failure. This can be fatal.

There is growing evidence of long-term effects including insomnia, depression and other psychological problems. It is possible that ecstasy can have an effect on normal brain activity even when the drug is no longer taken. Because the drug has stimulated so much release of serotonin, the cells cannot synthesise enough to meet demand once it has gone, leaving a feeling of depression.

Checkpoint

8.6 Construct a spider diagram showing the ways in which drugs can affect synaptic transmission. Include examples of the type of drugs that produce each effect.

8.8 Genes and the brain

Diseases of the brain and nervous system

Diseases associated with the nervous system are of major global importance. Mental illness is a term encompassing a wide range of diseases and disorders, and is the leading cause of illness and disability in the UK. It accounts for approximately 25% of the Government's total payments on sickness and disability. The sheer misery caused by conditions such as Huntington's disease, Alzheimer's disease, schizophrenia and bipolar disorder (manic depression) means that the search for effective treatments will be one of the major targets for twenty-first century biology.

Inheritance of diseases of the nervous system

Most disorders of the nervous system do not follow simple Mendelian rules of inheritance (described for monohybrid inheritance in Topic 2 and dihybrid inheritance in Topic 5). One example that does is the condition called Huntington's disease, a degenerative disease of the nervous system. Cells in the basal ganglia and cerebral cortex become damaged, leading to gradual physical and mental changes including involuntary movements, difficulty with speech, mood swings and depression. The phenotypic signs of the disease typically do not appear until the individual is 35–45 years old. The disease is fatal, and death occurs within 15–25 years of onset. The faulty gene responsible occurs on chromosome 4.

Q8.40 **a** From the pedigree diagram for a family with Huntington's disease shown in Figure 8.58, decide if the allele for Huntington's disease is dominant or recessive. Give a reason for your answer.
b Using the letters **H** and **h**, state the genotype for each of the 12 people in the diagram.

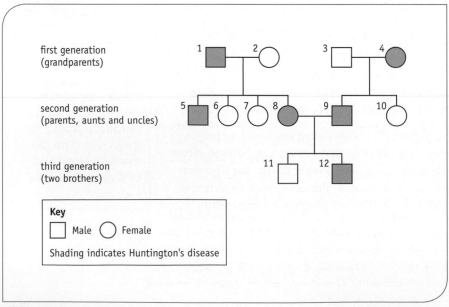

▲ **Figure 8.58** Family tree for Huntington's disease.

Polygenic inheritance of disease

Unlike Huntington's disease, most disorders of the nervous system are caused by the interaction of alleles at many loci. When a number of genes are involved in the inheritance of a characteristic, rather than just one, we call the pattern of inheritance **polygenic**. Conditions of the nervous system like Alzheimer's disease and schizophrenia are polygenic, as are a wide range of other conditions such as diabetes, coronary heart disease and some cancers.

In many human characteristics that show polygenic inheritance such as height and skin colour there are clearly additive effects of the alleles (see Key biological principle: Polygenic inheritance). With diseases, several genes may be implicated, but they simply confer a *susceptibility* to the condition, with environmental factors also contributing. Two people who inherit the same susceptibility genes may not both develop the illness; it will depend on environmental factors such as diet, exposure to toxins and stress. Such factors act as triggers to bring about the symptoms of disease. Conditions where several genetic and one or more environmental factors are involved are said to be **multifactorial**.

Twin studies allow determination of the genetic component of a disease. Identical twins, produced by division of the same fertilised egg, are genetically identical. The degree of similarity between the twins is a measure of the influence of the genes on that characteristic. 99% of identical twins have the same eye colour, and 95% the same fingerprint ridge count. Where the environment has a greater effect the similarity falls. One study found that if you have Alzheimer's disease and an identical twin sibling (sister or brother), your sibling has a 40% chance of having Alzheimer's disease. However, if you have Alzheimer's disease as a non-identical twin, your sibling has only a 10% chance of having Alzheimer's disease. This suggests that there is a significant but not inescapable genetic basis to Alzheimer's disease, with both genes and the environment influencing the development of the disease.

Key biological principle: Polygenic inheritance

In monohybrid and dihybrid inheritance, each locus is responsible for a different heritable feature. For example, one gene might be coding for flower colour and another for the shape of the petals. However, some characteristics are controlled by alleles at many loci.

In any introductory course on genetics it is common for eye colour to be used as an example of monohybrid inheritance – a single locus with brown dominant to blue. This is not entirely the case. Eye colour is an example of polygenic inheritance: alleles at several loci control eye colour. Eyes are brown or blue depending on the amount of pigment in the iris. Brown eyes have a great deal of pigment in the iris. This absorbs light, making the eye appear dark. Blue eyes have little pigment, so light reflects off the iris.

Let us say three loci are involved in the inheritance of this characteristic, each with alleles **B** and **b**. **B** adds pigment to the iris and **b** does not. If all three loci were homozygous for the allele **B**, the person's genotype would be **BB BB BB**. The additive effect would produce a dark brown iris, whereas **bb bb bb** would add no pigment to the iris, making it pale blue. A range of possible genotypes and phenotypes are possible between these two extremes, according to how many alleles add brown pigment, as shown in Table 8.7.

(Continued)

▼ **Table 8.7** Eye colour is a polygenic characteristic.

Number of alleles adding brown pigment	Example of genotype	Eye colour
6	BB BB BB	very dark brown
5	BB BB Bb	dark brown
4	BB BB bb	medium brown
3	BB Bb bb	light brown
2	BB bb bb	deep blue
1	Bb bb bb	medium blue
0	bb bb bb	pale blue

The greater the number of loci involved, the greater the number of possible shades.

If a pale-blue-eyed woman has children with a very dark-brown-eyed man they will have light-brown-eyed children as shown below.

	Mother	**Father**
Parents' phenotypes	pale blue	very dark brown
Parents' genotypes	**bb bb bb**	**BB BB BB**
Gametes	**bbb**	**BBB**
Offspring genotypes	**Bb Bb Bb**	
Offspring phenotype	light brown	

Q8.41 A deep-blue-eyed woman (**Bb bb bB**) has a child with light brown eyes. Her medium-blue-eyed partner (**Bb bb bb**) suspects that he is not the father. Copy and complete the Punnett square below and use it to explain that he could be the father.

	Mother	**Father**
Parents' phenotypes	deep blue	medium blue
Parents' genotypes	**Bb bb bB**	**Bb bb bb**
Possible gametes	**Bbb, BbB, bbb, bbB**	**Bbb, bbb**

		Gametes from father	
		Bbb	**bbb**
Gametes			
from mother			

Polygenic inheritance and continuous variation

Height, weight and skin pigmentation all involve polygenic inheritance. The combination of polygenic inheritance and environmental effects means that these characteristics show continuous variation (see Topic 3).

Each allele has a small effect on the characteristic, and the effects of several alleles combine to produce the phenotype of an individual. Suppose that only two genes were involved in the determination of height. The homozygous recessive genotype, **aabb**, might give a height of 20 cm above a baseline of 140 cm for adult women and 150 cm for adult men. In other words the recessive alleles (**a** and **b**) each contribute 5 cm to the height. The dominant alleles, **A** and **B**, each contribute 10 cm to the height – so the homozygous dominant **AABB** would give a height of 40 cm above our baseline. **AaBb** would add 30 cm to the height.

If two heterozygotes were crossed there would be a range of phenotypes as shown in Table 8.8.

▼ **Table 8.8** Range of phenotypes from crossing two **AaBb** heterozygotes.

Height above baseline/cm	20	25	30	35	40
Number of offspring with that height	1	4	6	4	1

Q8.42 **a** Sketch what a graph of the data in Table 8.8 would look like.
b What can you conclude about height from this graph?
c For the cross between two heterozygotes:
i state the parental phenotypes, parental genotypes, and possible gamete genotypes
ii draw up a Punnett square to show the possible offspring genotypes and phenotypes.

If, instead of two loci, there were several, the number of height phenotypes would increase. The more loci, the greater the number of height classes, and the smaller the differences between classes. If plotted this would produce a smooth bell-shaped curve typical of continuous variation (see Topic 3).

Traits such as height and skin colour are not simply the result of an individual's genotype. Their environment also has an influence. For example, diet has an effect on a person's height. A poor diet may prevent a person reaching their full height as predicted by their genotype. If you like, their genotype gives their *potential* height. This in combination with their environment determines their *actual* height. The result is that there are no clear height classes, but a continuous variation in height.

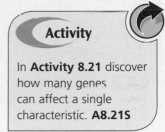

Activity

In **Activity 8.21** discover how many genes can affect a single characteristic. **A8.21S**

What can genetics tell us about the structure and activity of the brain?

With the sequencing of the human genome now complete, scientists are working to identify the function of every gene in the human genome. It has been estimated that about 50% of the 25 000 to 30 000 genes in the human genome are expressed in the nervous system. This high level of complexity makes understanding the structure and activity of the brain very challenging.

It is sometimes possible to find genes that are involved in brain disease and to see how gene expression is altered in the diseased tissue. This can provide insights into normal brain development and function – as well as helping in the development of better targeted drug treatments and other therapies.

Finding the gene

One technique used to find these disease genes is called **linkage analysis**. This follows the inheritance of DNA markers in affected families. Micro-satellites (see Topic 6) are now being used as markers. The locations of numerous micro-satellites are known, and by using restriction enzymes, PCR and gel electrophoresis it is possible to identify the markers. If a marker is consistently inherited with the disease (Figure 8.59), it suggests that the gene responsible for the disease is located on the chromosome close to the marker. The precise position of the gene can then be located by investigating that section of the DNA in more detail.

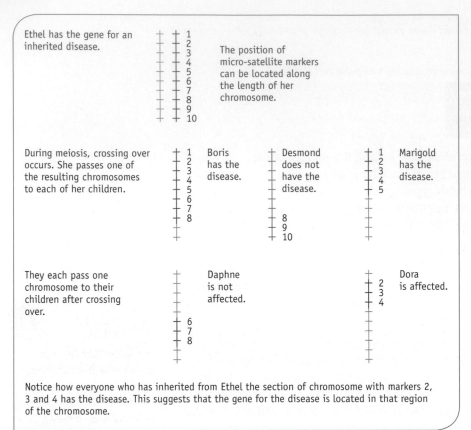

Ethel has the gene for an inherited disease.

The position of micro-satellite markers can be located along the length of her chromosome.

During meiosis, crossing over occurs. She passes one of the resulting chromosomes to each of her children.

Boris has the disease.

Desmond does not have the disease.

Marigold has the disease.

They each pass one chromosome to their children after crossing over.

Daphne is not affected.

Dora is affected.

Notice how everyone who has inherited from Ethel the section of chromosome with markers 2, 3 and 4 has the disease. This suggests that the gene for the disease is located in that region of the chromosome.

▲ **Figure 8.59** Linkage analysis can be used to help locate the position of a gene.

Inheritance of Alzheimer's disease

Alzheimer's disease is a form of progressive dementia. It is a multifactorial disease; age, genes and environment are all thought to be factors which increase the risk of developing the disease. Symptoms include confusion and memory loss.

Six genes have been identified as increasing susceptibility to Alzheimer's disease. In a few families, a genetic fault on chromosome 21 in the APP gene for production of the precursor of a particular protein (known as an amyloid protein) has been implicated in the development of Alzheimer's. Many cases of Alzheimer's have been linked to possession of alleles of a gene known as ApoE. There are three ApoE alleles (ApoE2, ApoE3 and ApoE4), and having two alleles of ApoE4 seems to increase the risk of getting the disease ten-fold. The ApoE gene controls production of a lipoprotein used in the repair of cell membranes in damaged neurones. The ApoE4 allele produces a variant of this, an amyloid protein that is deposited in insoluble plaques in the brain.

Unfortunately how the genes and environmental factors interact to cause Alzheimer's disease is still unclear. It is currently being debated whether many of the dementia conditions are single diseases triggered by a number of genes, or a number of diseases whose symptoms are very similar.

Looking at the expression of genes

Functional genomics

Functional genomics (the study of genes and their functions) has allowed researchers to look at the expression of more than 10 000 genes at any one time, helping us to make comparisons between diseased and normal tissue. If the expression of a gene is changed in a diseased cell compared with a normal cell, the gene may well have a role in the disease.

Comparing large numbers of genes is done using a microarray. This is a $1\,cm^2$ grid on a slide. Each of the squares on the grid (and there can be thousands of squares) has the DNA for a specific gene. When genes are expressed, they make mRNA transcripts. All the transcripts produced within a cell are tagged with fluorescent dye. These are then placed on the microarray where they bind to their corresponding genes. In this way it can be determined which genes are being expressed in the cell. Comparisons can be made between healthy and diseased tissue to identify any differences in gene expression.

Knock-out studies

Scientists can look at gene expression over time and this allows them to look at the development of the brain and how its structure and function change as we age.

Knock-out studies, where mice are bred with a particular gene switched off, are also used to investigate the role of individual genes in normal brain activity. For example, in studies on the sensory connections from mouse whiskers to the brain, mice have been bred with one gene switched off. This gene was thought to be involved in the organisation of the neurones in the brain into distinct groups, with each group connecting to a single whisker. With the gene switched off the neurones, as expected, failed to organise into groups. Humans don't have whiskers but other knock-out studies are in progress which should tell us about genes that have much the same role in mice and in ourselves, for example the gene involved in Huntington's disease.

We began Topic 8 with Kenge on the African plains mistaking buffalo for insects, and we have ended it by considering knock-out mice, a product of modern biotechnology. Understanding how the brain works requires a remarkable array of approaches which is why this topic contains such diverse subjects as optical illusions, twin studies, MRI scans, ecstasy use, animal experiments and learning. To a certain extent each of the topics in Salters-Nuffield Advanced Biology has covered a great range of material. Now that you have survived to the end of the course: congratulations! We hope you feel that it has all come together. We are confident that this approach will help you to understand current issues in biology whether you go on to study biology further or not. All the best!

Activity

Activity 8.22 lets you summarise the methods used to compare the contributions of nature and nurture to brain development discussed throughout the topic. **A8.22S**

Activity

Use **Activity 8.23** to check your notes using the topic summary provided. **A8.23S**

Topic test

Now that you have finished Topic 8, complete the end-of-topic test.

Answers to in-text questions

Topic 5

Q5.1 **a** A is a different species from B since 2% is a high percentage of genetic difference;
b A and C are either the same species or very closely related species as they can interbreed and produce fertile young;
c B and D are closely related but distinct species as the males that result from interbreeding are infertile;
d E and F are the same species since there is no genetic difference between them;

Q5.2 The assumptions made may be incorrect; e.g. the ratio of butterfly species to total insect species may vary widely across the globe; each tree species may not be a host to the same number of species;

Q5.3 **a** *Ranunculus*; **b** Meadowsweet and dropwort are more closely related; they both belong to the same genus;

Q5.4 *Chaetodon auriga*;

Q5.5 They have a long extended part to their dorsal fin;

Q5.6 Plants are easier to identify; a much great proportion of the world's plants than animals have been identified; plant diversity is related to animal diversity;

Q5.7 **a** Hotspots appear to be in the warmer regions of the world; greater availability of sunlight and precipitation provide the best conditions for plant growth; plant and associated animal diversity will tend to be lower where conditions are less favourable; areas more recently glaciated will have lower diversity; a number of the hotspots are mainly or completely made up of islands; being isolated makes speciation more likely;
b Species on islands easily become reproductively isolated; adaptive radiation can be a powerful force for speciation on islands; since only a small number of species colonise an area with many niches;

Q5.8

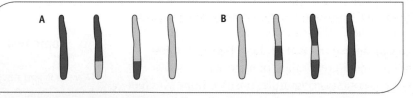

Q5.9 Black, short-haired; black, long-haired; brown, short-haired; brown, long-haired; 1:1:1:1;

Q5.10 **CCLL, CcLL, CCLl, CcLl**; all curly tail, large ears;

Q5.11 Equal numbers of **LLWw, LLww, LlWw, Llww**; half the offspring have large leaves with white flowers and half have large leaves with pink flowers, i.e. 1:1 ratio;

Q5.12 **a** Brine shrimps; algae; microorganisms; **b** Salt water; mineral substrate; air; light; temperature;

Q5.13 Just about all of them can be; for example, humans can affect the light that reaches a forest floor by cutting down trees; they can affect climate through the greenhouse effect; they can affect parasitism by using veterinary products; and so on;;;

Q5.14 Wave action; salinity; temperature; light; desiccation due to exposure at low tide; geology; topography; oxygen availability;

Q5.15 Being prickly; stinging; mimicry – the white deadnettle gets left alone because it looks like a nettle even though it doesn't sting; being unpalatable – the common field buttercup has an unpleasant taste and so cows leave it alone; the mat grass on moorland has silica in its leaves which make it rough so the sheep eat more tender grasses instead;

Q5.16 Large ears; large back legs; white tail;

Q5.17 Producer: *Laminaria*, *Chondus crispus*, toothed wrack, *Corallina*, bladder wrack, channel wrack, spiral wrack, lichens;
Primary consumer: flat periwinkle; limpet; rough periwinkle; black periwinkle;
Secondary consumer: dog whelk; starfish;
any appropriate adaptation for each example selected;;;

Q5.18 **a** H_2O; **b** Light; **c** Reduced NADP; **d** ATP;

Q5.19 **a** $H_2O \rightarrow 2H^+ + 2e^- + \frac{1}{2}O_2$; **b** $2H^+ + 2e^- + NADP \rightarrow$ reduced NADP;

Q5.20 **a** RuBP; **b** GALP; **c** 12 GP are needed to make 12 GALP, this provides the two GALP to make glucose and the 10 to recreate RuBP;

Q5.21 **a** Algae; **b** Brine shrimps; **c** Flamingo – this filters vast quantities of brine shrimps from the water using its specialised beak; **d** Eagle;

Q5.22 The blue; and red parts of the spectrum;

Q5.23 **a i** $23\,140 - 17\,820 = 5320\,kJ\ m^{-2}\ y^{-1}$;
ii $5100 - 1960 = 3140\,kJ\ m^{-2}\ y^{-1}$;
iii GPP = NPP + R = $26\,000 + 8000 = 34\,000\,kJ\ m^{-2}\ y^{-1}$;
GPP in the tropical rainforest is very high due to the high incident light, high temperatures and constant availability of water; the high temperature also means that the respiration rate is high and this means that NPP is not as high as the high rate of GPP might suggest;
GPP in the pine forest is much lower than the rainforest; partly because it is colder; however, although GPP is less than a quarter of the tropical rainforest GPP, the NPP is over half of the tropical rainforest NPP; because respiration is much less than in a rainforest;
Maize is very efficient at trapping light energy and so has a very high GPP; it has been bred to have a very high NPP;
b % efficiency = $\frac{34\,000}{2 \times 10^6} \times 100$; = $0.017 \times 100 = 1.7\%$;
c Most of the light is the wrong wavelength; and is not absorbed (by the chlorophyll/chloroplast); some is reflected by the leaves; some passes through the leaves without being absorbed; some warms the leaves; some is lost during the light-dependent reactions of photosynthesis;
d A desert has very low NPP due to lack of water; limiting photosynthesis; and sparse vegetation;

Q5.24 **a** $125\,kJ\ m^{-2}\ y^{-1}$; **b** 4%;

Q5.25 **a** 12.5%; **b** 20%;

Q5.26 **a** Due to variation in the population some individuals would have long necks; long necks give individuals an advantage, individuals with longer necks are better adapted to conditions in their environment, they survive and breed; some of the variation is heritable; their offspring are more likely to have a long neck so long necks become more common in the population; Today we would explain the process in terms of selection acting on alleles; Darwin and Wallace lacked our current knowledge of genetics;
b Acquiring a trait does not produce a change in the genotype; the DNA remains unaltered so the new characteristic cannot be passed on to the next generation;

Q5.27 1**A**; 2**B**; 3**C**; 4**D**; 5**E**; (3**B** and 2**C** are also acceptable;)

Q5.28 **a** Many biologists would argue that they are different subspecies but belong to the same species; **b** There are arguments for and against the control of ruddy ducks; control involves shooting the ducks; if you are in favour of conserving biodiversity you may favour control; if you are in favour of minimising animal suffering you may be against control;

Q5.29 *Arguments for*: role in academic research; captive breeding with careful use of studbooks to maintain genetic diversity; reintroduction programmes; education; *Arguments against*: animals are exhibited on the basis of their 'crowd-pulling' power, rather than on their endangered status; animals are kept in inappropriate conditions; both in terms of their physical environment (poor floors in cages, for example); and their mental and social well-being (limited feeding stimuli or atypical social grouping); capture of animals for exhibition may seriously deplete wild populations; reintroduction of species to the wild cannot be guaranteed;

Q5.30 **a** Scarcity of surface water; competition for these water bodies; high densities of aquatic predators; foam reduces risk of desiccation;
b Financial cost of flying someone out, their food and accommodation; logistics of working in the field raise problems such as finding the nests in the first place; avoiding disturbance to egg-laying and care of young, the impact of the weather on conducting regular observations; liaison with local people;

Q5.31 Robert; he is the only offspring from the two founders on the right; his parents are related so he is inbred but he can be paired with an unrelated animal;

Q5.32 *Advantages*: same environment as they will have to get used to after reintroduction; *Disadvantages*: risks of disease and death due to weather; parasites; predators;

Q5.33 **a** $(40 \div 6500) \times 9000 = 55\%$;
b Increased material standard of living;

Topic 6

Q6.1 Check for any forms of identification; e.g. diary, bank card, driving licence, etc.;

Q6.2

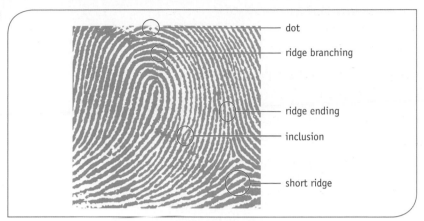

dot
ridge branching
ridge ending
inclusion
short ridge

Q6.3 Identification of dead bodies; identification of people who have lost their memory; laser scanning of fingerprints can be used for security identification;

Q6.4 Predict that the more closely related two people are; the more similar will be their fingerprints;

Q6.5 More similar fingerprints are obtained from closely related people than from unrelated ones;

Q6.6 Person 2;

Q6.7 agta; 6;

Q6.8 The size of the DNA fragments; shorter fragments go further;

Q6.9

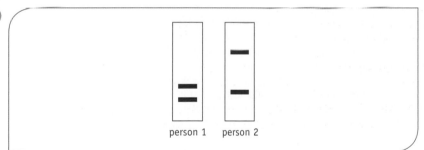

person 1   person 2

Q6.10 **a** Poppy; for each satellite the chick has inherited one copy of the repeat sequences from Poppy; one copy of A could come from Patsy but not B or C;
b Both children share some bands with each parent so the father is their biological parent; (The mother's bands are marked in red and the father's in blue.)

Q6.11 The more satellites are used; the more confident; one can be about the conclusions drawn;

Q6.12 At elevated external temperatures such as found in the tropics; there will be little cooling of the body;

Q6.13 Larger mass; sitting upright; less air movement; low humidity; clothing;

Q6.14 The body is likely to have cooled to the surrounding temperate; the body was found in May; when the surrounding temperature would be about 18°C; so cannot tell if the person died in the last 24 hours;

Q6.15 **a** Speeds up; raises the body temperature so rate of enzyme action increases;
b Speeds up; allows entry of bacteria;
c Slows down; destroys the bacteria and denatures enzymes involved in decomposition;

Q6.16 Between 36 and 72 hours ago;

Q6.17 Look at the characteristic features of the insect and if not immediately recognisable use a key to identify the species;

Q6.18 4 days;

Q6.19 Approximately 30 hours;

Q6.20 Eggs laid between 24 and 28 June; earliest date of death 22 June assuming that the eggs were laid within two days of death;

Q6.21 **a** Higher temperatures would speed up succession; more rapid development of the organisms present; due to faster enzyme action; conditions on the body change more quickly allowing entry of the next species in the succession; not true if temperature greater than about 40 °C;
b Humidity; soil conditions; pH; chemical pollution;

Q6.22 **a** Atherosclerosis;
b Genetic predisposition; diet high in cholesterol and LDL lipids; smoking; high alcohol consumption;
c Aneurysms are prone to rupture (when they reach about 6–7 cm in size); the resulting blood loss from the circulatory system and the resultant shock can be fatal;

Q6.23 a; b; c; d; f;

Q6.24 **a** virus; **b** bacteria; **c** virus; **d** bacterium; **e** virus;

Q6.25 Some people do not consider viruses to be truly living organisms because they are not capable of independent reproduction or energy use; however they do have characteristics of living organisms; because they can reproduce; pass genetic information to the next generation; and evolve;

Q6.26 The *Mycobacterium* is 22 times larger than the bacteriophage;

Mycobacterium length: 27 mm

Actual length $= \dfrac{27\,\text{mm}}{5200}$

$\qquad\qquad = 0.00519\,\text{mm or } 5.19\,\mu\text{m}$

Bacteriophage length: 4 mm

Actual length $= \dfrac{4\,\text{mm}}{17\,000}$

$\qquad\qquad = 0.000\,24\,\text{mm or } 0.24\,\mu\text{m}$

Q6.27 Barriers; between maternal and fetal blood;

Q6.28 By hydrolysing the polysaccharide;

Q6.29 Increased blood flow to the infected area makes it red; it is hot due to increased metabolic activity; it is swollen due to the leakage of fluid from capillaries into the tissue;

Q6.30 A;

Q6.31 Antibody production starts sooner; the speed of production is faster; and a higher concentration is produced; so the immune system can deal with the microbe before symptoms appear;

Q6.32 The conditions inside tubercules are anaerobic so there is no oxygen available for the bacteria to use in respiration;

Q6.33 **a** Less gas exchange; the lung surface area is reduced so gas exchange is less efficient;
b Breathing rate increases; to make up for the loss of gas exchange surface;

Q6.34 At high temperatures proteins in the cell are denatured; metabolism is disrupted as enzymes denature; membranes are disrupted;

Q6.35 Hair erector muscles; muscles in the walls of the arterioles in the skin; voluntary muscles;

Q6.36 Better diet (plenty of fruit and vegetables); reduced alcohol intake; stop smoking; better housing conditions;

Q6.37 TCCAAT;

Q6.38 **a** Passive artificial; passive natural; **b** Active artificial; active natural; **c** Passive artificial; passive natural; **d** Active artificial; active natural; **e** Active artificial; active natural; **f** Active artificial; passive artificial;

Q6.39 **a** 1991; **b** Before 1968 each measles epidemic greatly reduced the number of potential new hosts; after a couple of years enough new babies had been born; to allow a new epidemic; between 1968 and 1988 the proportion of the population capable of contracting measles was falling; but herd immunity had not yet been attained;

Q6.40 If penicillin is an enzyme it will denature and lose its antibacterial effect when heated; heating should not affect a toxic chemical; so heat penicillin and then place it on agar plates inoculated with bacteria; observe results; draw valid conclusion;

Q6.41 Option 1: *Disadvantages*: Previous patients may not have the same severity of TB as the next 100 patients that come into the hospital; previous treatments may have been so bad that anything could be better; *Advantages*: The doctor doesn't have the ethical dilemma of choosing which patients get the treatment and which don't;
Option 2: This would select a random sample but it is not a blind trial so the doctor may bias the experiment because she/he knows who is getting the treatment;
Option 3: This is really the best option if the doctor does not select the patients; in many such trials, the test is double-blind so that neither the doctor nor the patient knows whether the patient is getting the new drug or an inert substance called a placebo; this reduces the chances of the patient or the doctor exaggerating any improvements they see and deciding that these are the result of the new drug; (When the effect of streptomycin was tested by the Medical Research Council in the 1940s, they decided on option 1 for patients with tuberculous meningitis, because it was 100% fatal with the existing therapy. However, for pulmonary tuberculosis, they did a randomised trial (option 3).)

Q6.42 6;

Q6.43 Human cells do not have cell walls;

Q6.44 **a** They can develop a drug with a complementary shape which binds to the pump; stopping it functioning;
b If the pumps occurred in humans they might pump the antibiotic out of the cell; making it ineffective against any bacteria in the cell;

Q6.45 Avoid the risky behaviour that resulted in HIV infection; vaccination might have protected her from TB; avoiding overcrowded hostels where there may have been a higher risk of TB infection;

Topic 7

Q7.1 A;

Q7.2 **a** Strength, flexibility/elasticity; **b** Synovial fluid; cartilage;

Q7.3 **a** A ligament; B cartilage; C cartilage/pad of cartilage;
b Tendon; muscle; synovial membrane; synovial fluid; fibrous capsule;

Q7.4 It would take too long; to move proteins synthesised from mRNA from a single nucleus; to reach the furthest parts of the cell;

Q7.5 5;

Q7.6 The central band disappears; because the actin has moved and there is now no point at which there is only myosin on its own;

Q7.7 Better insulation reduces heat loss; so less energy is used to maintain core body temperature;

Q7.8 No oxygen is used;

Q7.9 Oxidation–reduction reactions occur as the electrons pass along the electron transport chain; the final electron acceptor is oxygen; phosphate is added to ADP to form ATP;

Q7.10 **a** 4; 2ATP net from glycolysis and 2ATP from Krebs cycle (one for each pyruvate molecule entering from glycolysis);
b 34; each reduced NAD produced results in 3ATP; each reduced FAD produced results in 2ATP; 2 reduced NAD in glycolysis, 4 reduced NAD and 1 reduced FAD for each turn of the Krebs cycle making 8 reduced NAD and 2 reduced FAD for each glucose molecule;

Q7.11 It prevents the cell overheating; allows the controlled release of energy in small useful quantities;

Q7.12 To maintain rapid blood flow through muscles to supply oxygen; and remove lactate;

Q7.13 The untrained athlete; larger shaded area on Figure 7.31; they take up and transport oxygen more slowly so take longer to reach maximum oxygen uptake; their period of anaerobic respiration is longer;

Q7.14 **a** B; **b** A; **c** C;

Q7.15 **a** Glycolysis; and ATP/PC; **b** Aerobic;

Q7.16 It converts fats and proteins to glycogen; for storage in muscles and liver;

Q7.17 **a** Cardiac output = SV × HR; = 75 × 70; 5250 cm^3 per minute or 5.25 dm^3 per minute;

diac output/HR; = $\frac{5250}{50}$; 105 cm^3;

HR: = $\frac{5250}{33}$; 159 cm^3;

Q7.18 **a** If pressing on the neck causes increased blood pressure in the carotid artery; blood pressure sensors in the carotid artery would signal to the cardiovascular control centre; which in turn would stimulate the vagus nerve; reducing heart rate and thus pulse measurement;
b The wrist; or groin;

Q7.19 Anticipatory rise due to the effect of adrenaline on the heart; increases oxygen supply to the muscles in preparation for the activity about to occur;

Q7.20 $6 \, dm^3 \, min^{-1}$;

Q7.21 **a** $0.65 \, dm^3$; **b** $2.7 \, dm^3$; **c** Tidal volume × rate of breathing; $= 0.65 \times 16 = 10.4 \, dm^3$ per minute; **d** Rate of oxygen consumption = volume of oxygen used (dm^3)/time (s); $= 0.5/20$; $= 0.025 \, dm^3 \, s^{-1}$ or $25 \, cm^3 \, s^{-1}$;

Q7.22 The depth and rate of breathing increase so there is a greater volume of air inhaled and mixed with the residual air in the lungs; so concentration of oxygen increases; the higher concentration makes the diffusion gradient between the alveolar air and the blood steeper; increasing the speed of gas exchange; an advantage given the raised metabolic rate;

Q7.23 Stretch receptors signal the start of movement; allowing ventilation to increase before there is a build-up of the waste products of respiration;

Q7.24 The increased concentration of oxygen in the blood; detected by chemoreceptors; slows breathing rate and decreases depth of breathing;

Q7.25 **a** Aerobic respiration occurs within the mitochondria; large numbers allow slow-twitch fibres to have a greater capacity for aerobic respiration;
b Calcium ions released from the sarcoplasmic reticulum initiate muscle contraction; more sarcoplasmic reticulum allows rapid, repeated contraction of the muscle;
c Myoglobin stores oxygen within the cells for use in aerobic respiration;
d Slow-twitch via aerobic respiration; fast-twitch using anaerobic glycolysis reactions;
e Fast-twitch; poor supply of oxygen to the fibre; uses anaerobic respiration; rapid build-up of lactate;

Q7.26 **a** Fast-twitch; **b** Slow-twitch;

Q7.27 **a** 37–8 °C; **b** Low temperatures lead to low metabolic rates as the enzyme-controlled reactions slow; high temperatures increase the rate of metabolic reactions initially; but then it declines as the higher temperature denatures the enzymes;

Q7.28 **a** Increases heat energy loss by conduction; and evaporation;
b Less energy is lost by evaporation so it is harder for athletes to keep their body temperatures down to a safe level;

Q7.29 Evaporation from gas exchange surfaces; lowering hairs to increase energy loss by convection; conduction; and radiation;

Q7.30 **a** Conduction; **b** Some energy is transferred to the body cells as a waste product of normal metabolism; shivering increases resting metabolism 3- to 5-fold; transferring additional energy to the body cells; nerve impulses to the arterioles in the skin cause vasoconstriction; resulting in restricted blood flow through the skin; this reduces energy loss by radiation, conduction and convection; hair raising is pretty ineffectual in humans, particularly for cross-Channel swimmers who coat themselves in protective grease (petroleum jelly);

Q7.31 Reduced risk of upper respiratory tract infections with moderate volume or moderate intensity exercise; increased risk of infection with large amounts of or high intensity exercise;

Q7.32 **a** B cells; **b** Inflammation; phagocytosis; antimicrobial proteins; **c** Few T helper cells so less cytokines produced; T killer cells will not be activated by the cytokines;

Q7.33 Moderate exercise increases the number of natural killer cells; intense exercise reduces the number and activity of natural killer cells, phagocytes, B cells and T helper cells;

Q7.34 It would increase the number of red blood cells and hence the amount of haemoglobin; improving the blood's oxygen-carrying capacity; enhancing oxygen delivery to muscle tissue and hence improving aerobic capacity;

Q7.35 Blood clots in the arteries or veins;

Q7.36 Sprint events rely on anaerobic respiration so performance is not dependent on the athlete's aerobic capacity;

Q7.37 To help determine if athletes are taking EPO as a performance-enhancing drug;

Q7.38 They are lipid based and so dissolve;

Topic 8

Q8.1 They produce rapid responses; important for protection and survival;

Q8.2 Other neurones must be involved; relay neurones within the central nervous system; motor neurones to muscles in the arm; which are under conscious control;

Q8.3 The one on the left;

Q8.4 Radial;

Q8.5 **a** Rods and cones in the retina; **b** Optic nerve; **c** Brain; **d** Iris muscle;

Q8.6 To prevent damage to the retina from high-intensity light; in dim light it ensures maximum light reaches the retina;

Q8.7 To protect the eye from sudden flashes of bright light;

Q8.8 Three;

Q8.9 The channel's opening is dependent on changes in voltage;

Q8.10 No (unless ATP was added); the polarisation of the membrane is maintained by the concentration gradients achieved by the action of energy-requiring sodium–potassium pumps; membrane integrity is lost in a dead axon;

Q8.11 A new action potential will only be generated at the leading edge of the previous one; because the membrane behind it will be recovering/incapable of transmitting an impulse; the membrane has to be repolarised and return to resting potential before another action potential can be generated;

Q8.12 Site of production: produced in the testes by males and in small amounts by the adrenal glands in both males and females;
Method of transport: in the blood;
~~~ of target cells: through out the body including male sex organs, skin
~~~ any cells involved in the development of the secondary sexual

Effect on the target cells: binds to androgen receptors on target cells modifying gene expression to alter the development of the cell; for example, increasing anabolic reactions such as protein synthesis in muscle cells, increasing the size and strength of the muscle;

Q8.13 Might expect photoreceptors to be on the surface of the retina, but they form a deeper layer/light has to travel through the other layers including blood vessels before reaching the photoreceptors;

Q8.14 **a i** Active transport; **ii** Diffusion;
b One that lets any positive ions through, such as Na^+ and Ca^{2+};
c Because sodium ions are being actively transported out; and their re-entry through ion channels is prevented; increasing the potential difference across the membrane;

Q8.15 The region of the brain concerned with vision processing is the occipital lobe which sits at the back of the cortex and is closest to the back of the head (thus a blow to this area would cause a disturbance in vision);

Q8.16 Frontal lobe; parietal lobe; motor cortex; cerebellum;

Q8.17 Parietal lobe/basal ganglia;

Q8.18 There are many conditions that can cause brain damage, including lack of oxygen, carbon monoxide poisoning, toxic chemical exposures, infectious diseases, tumours, strokes and genetic conditions;;;

Q8.19 Cerebellum; motor cortex in frontal lobe;

Q8.20 **a i** fMRI; **ii** MRI; **b** The posterior hippocampus is involved in remembering detailed mental maps;

Q8.21 Visual stimulation with both light and patterns;

Q8.22 Because kittens are born blind, early deprivation (under three weeks) would have no effect; by three months connections to the brain have been made, and deprivation has no effect/the critical period has ended; the critical period is at about four weeks of age so lack of stimulation from the kitten's environment at this time severely affects visual development;

Q8.23 For some birds their song seems to be innate, due to nature and not nurture; for others both nature and nurture seem to be required;

Q8.24 1; 5; 3; 2; 4 or 1; 5; 3; 4; 2;

Q8.25 Strawberry; pencil;

Q8.26 **a** Bart Simpson; **b** Marilyn Monroe; **c** Michael Reiss (Director of the SNAB project); Anyone who could not be recognised has probably never been seen before;

Q8.27 Synapses from the optic nerve axons to the visual cortex will have been weakened or eliminated; Binocular cells in the visual cortex may only receive sensory information from one eye; so they will not have two views to compare, making stereoscopic vision difficult or impossible;

Q8.28 The fact that there is a correlation between two variables does not necessarily mean that there is a causal relationship;

Q8.29 Familiarity with the relative sizes of the elephant and the antelope; the elephant in the picture is smaller than the antelope and man so it is in the background; overlap, one hill partly hiding another tells us it is closer;

Q8.30 Some apparent cross-cultural differences in perception may occur because people cannot report perceptual differences; there may, for instance, be a language difficulty; researchers from Western societies chose the tests; there has been a strong emphasis on two-dimensional visual illusions which may favour subjects from the carpentered world;

Q8.31 She will show symptoms of distress or fear, and may refuse or cry;

Q8.32 **a** The baby has not experienced this before so cannot have learned it;
b Some perceptual development has certainly taken place since birth; the experiment requires the baby to crawl and this isn't possible for several months;

Q8.33 The fact that these animals had had little time for learning suggests that the behaviour is innate;

Q8.34 Kenge was not used to vast open spaces and would have had little opportunity to look far into the distance; with no experience of seeing distant objects he would have had little experience of certain environmental depth cues such as *size constancy* – the tendency to perceive the same object as always being the same size, however far away it is; as a result he saw the buffalo simply as very small animals;

Q8.35 **a** Neutral stimulus – sound; unconditioned stimulus – receiving food; conditioned response – salivation;
b Neutral stimulus – sound; unconditioned stimulus – puff of air; conditioned response – blinking;

Q8.36 Parietal lobe; temporal lobe; hippocampus;

Q8.37 **a** The neurones are not myelinated; **b** It has large diameter axons;

Q8.38

Q8.39 Breast cancer is common in women; over time hundreds of thousands of women die from it; each woman suffers more than each mouse; friends and relatives of a woman who dies young also suffer emotional pain;

Q8.40 **a** The allele is dominant; individuals 8 and 9 who both have the condition produce an offspring who does not; individuals 8 and 9 must be heterozygous;
b H = Huntington's disease, **h** = non Huntington's disease; 1 **Hh**; 2 **hh**; 3 **hh**; 4 **Hh**; 5 **Hh**; 6 **hh**; 7 **hh**; 8 **Hh**; 9 **Hh**; 10 **hh**; 11 **hh**; 12 **Hh** or **HH**;

Q8.41

	Bbb	**bbb**
Bbb	BB bb bb (2)	Bb bb bb (1)
BbB	BB bb Bb (3)	Bb bb Bb (2)
bbb	Bb bb bb (1)	bb bb bb (0)
	Bb bb Bb (2)	bb bb Bb (1)

The number of alleles adding pigment is shown in brackets. The child with three of these alleles will have light brown eyes. There is a 1 in 8 chance of the couple having a child with brown eyes;;;;

Q8.42 a

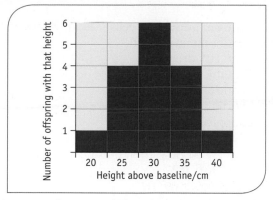

b Height shows continuous variation;

c i

	Mother and father
Parental phenotypes	Baseline height plus 30 cm
Parental genotypes	**AaBb**
Possible gametes	**AB Ab aB ab**

ii

	AB	**Ab**	**aB**	**ab**
AB	**AABB** 40	**AABb** 35	**AaBB** 35	**AaBb** 30
Ab	**AABb** 35	**AAbb** 30	**AaBb** 30	**Aabb** 25
aB	**AaBB** 35	**AABb** 30	**aaBB** 30	**aaBb** 25
ab	**AaBb** 30	**Aabb** 25	**aaBb** 25	**aabb** 20

A and **B** contribute 10 cm above baseline height while **a** and **b** add 5 cm to baseline height; The ratio of heights above baseline (cm) 20, 25, 30, 35, 40 in the offspring is 1:4:6:4:1;

Index

Acknowledgements

The authors and publishers would like to thank the following for permission to use copyright material: Biodiversity quote, page 4 reproduced by permission of the Secretariat of the Convention on Biological Diversity; Fig 5.5, p4 adapted from EO Wilson (1992) *The Diversity of Life*, London: Allen Lane. Reproduced by permission of Penguin Books Ltd; Fig 5.10, p10 reprinted from the original in EB Klemm *et al.* (1995) *The Living Ocean*, Honolulu: CRDG, University of Hawaii. © University of Hawaii. Reprinted with permission from the Curriculum Research & Development Group[†]; Fig 5.16, p16 © Nature, Vol. 403, 24 February 2000, reproduced with permission; Fig 5.25, p24 and Fig 5.37, p39 adapted from S Tomkins and M Dockery (2000) *Brine Shrimp Ecology*, London: British Ecological Society; Data in Fig 5.42, p42 reproduced with permission of the Nuffield Foundation; Table 5.5, p55 reproduced with permission from the UK Biodiversity Action Plan; Fig 5.69, p63 adapted from the GLTCP website (http://www.hrw.com/science/si-science/biology/animals/glt/); Fig 6.17, p82 and Tables 6.5 and 6.6, p85 adapted from KGV Smith (1986) *A Manual of Forensic Entomology*, London: Trustees of the British Museum (Natural History); Fig 6.20, p84 adapted from article 'On maggots and murders: forensic entomology' by Martin Hall on the Natural History Museum website (http://www.nhm.ac.uk); M Lee Goff quote, p84 reprinted by permission of the publisher from *A Fly for the Prosecution: How Insect Evidence Helps Solve Crimes* by M Lee Goff, p14, Cambridge, Mass: Harvard University Press. © 2000 by the President and Fellows of Harvard College; Fig 6.34, p95 adapted from Fig 43.4, p842 in N Campbell, J Reece and L Mitchell (1999) *Biology*, 5th Edn © 1999 by Benjamin/Cummings, an imprint of Addison Wesley Longman, Inc. Reprinted by permission of Pearson Education, Inc; Table 6.7, p104 adapted from Fact sheet 104 'Tuberculosis', revised April 2005, on the World Health Organisation website (http://www.who.int); Fig 6.47, p112 reproduced by kind permission of the Joint United Nations Programme on HIV/AIDS (http://www.unaids.org); David Livermore quote, p129 reproduced with permission from article 'Setback in superbug battle: Patient shows resistance to new class of antibiotic', Sarah Boseley, *The Guardian*, 7 December 2002 © Guardian Newspapers Ltd 2002; Figs 7.31 and 7.32, p153 and Fig 7.43, p166 adapted from W McArdle, F Katch and V Katch (2005) *Essentials of Exercise Physiology*, Philadelphia: Lippincott, Williams and Wilkins; Fig 7.47, p170 from LT Mackinnon (2000) 'Special feature for the Olympics: effects of exercise on the immune system: overtraining effects on immunity and performance in athletes', *Immunol Cell Biol* **78**: 502–9; Table 8.2, p189 from AL Hodgkin (1958) 'Ionic movements and electrical activity in giant nerve fibers' *Proc R Soc Lond B* **148**. 1–37. All Crown Copyright material is reproduced with the permission of the Controller of HMSO and the Queen's Printer for Scotland.

[†]The Hawaii Marine Science Studies (HMSS) program is a one-year multidisciplinary course set in a marine context for students in grades 9–12. There are two companion student books, *The Fluid Earth* and *The Living Ocean*, which explore the physics, chemistry, biology, and geology of the oceans and their applications in ocean engineering and related technologies. HMSS is a product of the Curriculum Research & Development Group (CRDG) of the University of Hawaii.

Every effort has been made to contact copyright holders of material reproduced in this book. Any omissions will be rectified in subsequent printings if notice is given to the publishers.

The authors and publishers would like to thank the following for permission to use photographs:

T = top, **B** = bottom, **L** = left, **R** = right, **M** = middle

SPL = Science Photo Library

Cover: Corbis and Digital Vision

Page 2, **T** Alamy Images/Jack Sullivan; 2, **B** Durrell Wildlife Conservation Trust/Gerado Garcia; 3, **T** SPL; 3, **B** NHPA; 5, NHPA; 6, **ML** Nature PL; 6 **MR** Alamy; 8, **T** SPL; 8 **B** NHPA; 10, OSF; 11, Ardea x6; 17, Corbis; 28, FLPA; 29, FLPA; 30, SPL; 33, SPL; 41, OSF; 43, Galapagos Trust/Roving Tortoise; 45, iStockPhoto.com/Steve Geer; 46, SPL; 49, **T** Ardea; 49, **B** Flowerphotos; 50, **ML** OSF; 50, **MR** FLPA; 52, Alamy; 53, Durrell Wildlife Conservation Trust/Gerado Garcia; 54, SPL; 55, OSF; 56, Nature PL; 57 David Slingsby x2; 58, **T** Corbis; 58, **B** Corbis; 59, Corbis; 60, FLPA x2; 61, Durrell Wildlife Conservation Trust/Herpatology Dept x2; 63, FLPA; 64, Durrell Wildlife Conservation Trust/Philip Duffy; 65, NHPA; 68, **T** iStockPhoto.com/Darrell Fraser; 68, **B** SPL; 71, **T** SPL; 71 **BL** SPL; 71 **BR** Corbis/Muriel Dovic; 78, SPL; 82, SPL; 84, SPL; 86, SPL; 87, SPL; 88, SPL; 90, SPL; 92, SPL x2; 94, SPL; 103, SPL; 106, SPL; 111, SPL; 113, SPL; 121, Wellcome Medical Photo Library; 122, SPL; 124, SPL; 125 SPL; 126, SPL; 127, SPL; 130, **T** Photos.com; 130, **M** Getty Images UK/Photodisc; 130, **B** FLPA; 131, **T** Corbis; 131, **B** Action Plus; 132, Empics; 135, SPL; 136, SPL; 141, SPL; 146, SPL; 149, Empics; 150, Empics; 151, SPL; 152, Empics; 159, Andrew Rankin; 164, SPL; 165, Empics; 168, PA Photos; 169, Empics; 172, SPL x2; 173, **ML** Empics; 173, **MR** Corbis; 175, Mary Evans Picture Library; 176, Empics; 178, Corbis; 179, Empics; 180, **T** Digital Vision; 180, **B** Alamy; 182, SPL; 185, SPL x2; 187, SPL; 203, SPL; 207, SPL; 211, SPL; 212, SPL; 213, SPL; 216, SPL; 218, Florence Low; 220, **TL** iStockPhoto.com/Naomi Hasegawa; 220, **TR** SPL; 220, **BL** Source Unknown; 220, **BM** Corbis; 220, **BR** Michael Reiss; 221, Corbis; 224, Source Unknown; 229, **T** Alamy; 229, **B** SPL; 232, Corbis.